T0156095

Lecture Notes in Business Information Processing 507

LNBIP reports state-of-the-art results in areas related to business information systems and industrial application software development – timely, at a high level, and in both printed and electronic form.

The type of material published includes

- Proceedings (published in time for the respective event)
- Postproceedings (consisting of thoroughly revised and/or extended final papers)
- Other edited monographs (such as, for example, project reports or invited volumes)
- Tutorials (coherently integrated collections of lectures given at advanced courses, seminars, schools, etc.)
- Award-winning or exceptional theses

LNBIP is abstracted/indexed in DBLP, EI and Scopus. LNBIP volumes are also submitted for the inclusion in ISI Proceedings.

Shey-Huei Sheu
Editor

Industrial Engineering and Applications – Europe

11th International Conference, ICIEA-EU 2024
Nice, France, January 10–12, 2024
Revised Selected Papers

Editor
Shey-Huei Sheu
Asia University
Taichung, Taiwan

ISSN 1865-1348 ISSN 1865-1356 (electronic)
Lecture Notes in Business Information Processing
ISBN 978-3-031-58112-0 ISBN 978-3-031-58113-7 (eBook)
https://doi.org/10.1007/978-3-031-58113-7

This Springer imprint is published by the registered company Springer Nature Switzerland AG
The registered company address is: Gewerbestrasse 11, 6330 Cham, Switzerland

Preface

On behalf of the conference organizing committee, we would like to express our welcome to all the attendees to our conference, The 11th International Conference on Industrial Engineering and Applications (Europe) (ICIEA 2024-Europe), which was held in Nice, France during January 10–12, and focused on the latest achievements in industrial engineering and its applications.

ICIEA is the leading conference to disseminate, to all branches of industries, information on the most recent and relevant research, theories and practices in industrial engineering and applications. All submissions were subjected to rigorous review before an acceptance decision was made. This conference was attended by more than 1000 participants from 50 countries since 2014 and was successfully held in major cities such as Sydney, Singapore, Hong Kong, Nagoya, and Paris.

There were two plenary speakers and two keynote speakers invited by the conference.

Luiz Moutinho, University of Suffolk, UK; The Marketing School, Portugal; University of the South Pacific, Fiji
Okyay Kaynak, Bogazici University and Turkish Academy of Sciences, Turkey
Cristian Mahulea, University of Zaragoza, Spain
Maurizio Faccio, University of Padova, Italy

The conference received 90 submissions from the UK, the USA, Germany, Sweden, France, Portugal, Italy, Japan, Saudi Arabia, China, Philippines, India, Indonesia, South Korea, Colombia, Morocco, Turkey, Croatia, Thailand, etc. After double-blind peer review, with each submission receiving an average of two reviews, 16 full and 3 short papers were accepted for publication in these conference proceedings, and they are divided into 5 chapters; their topics are (1) Industrial Production Control and Management System, (2) Control Models and System Analysis in Industrial Engineering, (3) Industrial System Layout and Site Selection Planning, (4) Image-based Industrial Equipment Detection and Status Recognition, (5) Market-Oriented Industrial Production and Product Pricing Strategies.

We would like to express our sincere appreciation for the help of members of the conference committee, reviewers, and workers. Thanks to their work, our conference could be held successfully. We hope that you can benefit from this conference.

February 2024 Shey-Huei Sheu

Organization

International Advisory Chair

R. Sabherwal	University of Arkansas, USA

General Chairs

Roel Leus	KU Leuven, Belgium
Luiz Moutinho	University of Suffolk, UK

Program Chairs

Adrian E. Coronado Mondragon	Royal Holloway University of London, UK
Dimitri Lefebvre	Le Havre University Normandy, France
Roberto Montemanni	University of Modena and Reggio Emilia, Italy
Shey-Huei Sheu	Asia University, Taiwan
Juan Carlos Quiroz Flores	Universidad de Lima, Peru

Workshop Chair

Stella Sofianopoulou	University of Piraeus, Greece

Special Session Chair

Ghassen Cherif	Université Toulouse III - Paul Sabatier, France

Publicity Chairs

Sadok Turki	University of Lorraine, France
Amr Eltawil	Egypt-Japan University of Science and Technology, Egypt
Hazem W. Marar	Princess Sumaya University for Technology, Jordan

International Technical Committee

Ana Luísa Ramos	University of Aveiro, Portugal
Ramin Yahyapour	Georg-August-University Göttingen/GWDG, Germany
Yung Ting	Chung Yuan Christian University, Taiwan
Lucian-Ionel Cioca	"Lucian Blaga" University of Sibiu, Romania
Pongchanun Luangpaiboon	Thammasat University, Thailand
Shabir Hussain Khahro	Prince Sultan University, Saudi Arabia
Ronaldo Polancos	De La Salle University, Philippines
Lei Yue	Guangzhou University, China
Sirang Klankamsorn	Kasetsart University, Thailand
Zubair Ahmed Memon	Prince Sultan University, Saudi Arabia
Shaligram Pokharel	Qatar University, Qatar
Jorge Juliao	Universidade Católica Portuguesa, Portugal
Rabah Ammour	Aix-Marseille University, France
João M. G. Dias Alves Lourenço	Setúbal Polytechnique Institute - ESTSetubal, Portugal
M. Lionel Amodeo	University of Technology of Troyes, France
Josephine German	Mapua University, Philippines
Hao Yu	UiT The Arctic University of Norway, Norway
Abdelkhalak El Hami	INSA Rouen, Normandy University, France
Abdessamad Kobi	University of Angers, France
Carina Pimentel	University of Minho, Portugal
Jose Martin Diaz	Corporación Universitaria Minuto de Dios, Colombia
Lerdlekha Sriratana	Ramkhamhaeng University, Thailand
Mohammed Alkahtani	King Saud University, Saudi Arabia
Hugo d'Albert	Technical University of Munich, Germany
Ronaldo Polancos	De La Salle University, Philippines
Gitae Kim	Hanbat National University, South Korea
Nicolas Ragot	CESI LINEACT, France
Nuno Costa	Instituto Politécnico de Setúbal - ESTSetubal, Portugal
Dong-Hee Lee	Sungkyunkwan University, South Korea
Dominik Saubke	Helmut-Schmidt-Universität, Germany
Iris Ann Martinez	University of the Philippines Diliman, Philippines
Sara Perotti	Politecnico di Milano, Italy
Maurizio Faccio	University of Padova, Italy
Onur Golbasi	Middle East Technical University, Turkey
Manoharan R.	VIT University, India

Contents

Cost-Optimization of Condition-Based Maintenance Policies for a Two-Component Machine System with General Repairs and Process Rejects

Kevin Kenneth Kwan, Simon Anthony Lorenzo, and Iris Ann Martinez(✉)

Department of Industrial Engineering and Operations Research, University of the Philippines Diliman, Quezon City, Philippines
{sdlorenzo,igmartinez}@up.edu.ph

Abstract. This study discusses how to optimize the cost for condition-based maintenance, specifically in the case of a two-component machine system. This is important because, in manufacturing systems that are highly dependent on machines, the total cost of maintaining stochastically degrading components can be high (i.e., as a result of inspections, general repair including replacement, and also the cost of lost production and process rejects due to machine breakdowns). This study hypothesizes that the minimum total cost of a condition-based maintenance policy for a two-component series system can be found by jointly solving for (a) the optimal inspection interval, (b) the general repair conditions for the two components, and (c) the cost estimation of process rejects and breakdown during every maintenance inspection and maintenance performed; doing all of these will give a realistic cost structure as encountered in real-life manufacturing systems. Given this hypothesis, this study models the two-component system by setting up the operating states and the breakdown states and defining the system variables. Then, this study establishes the decision variables, i.e., for the inspection state and the maintenance states, and thereafter performs the exhaustive Monte Carlo simulation approach to get the optimal variables. Finally, this study uses the optimum decision variables to come up with policies for condition-based maintenance. To prove the validity of this hypothesis, a case study is performed on a two-component process of a cutter-grinder assembly of a local fast-moving consumer goods company in the Philippines. By applying the approach proposed by this study that integrates stochastic degradation, frequency of inspections, process rejects and general repair scenarios, the total cost of annual maintenance is seen to be reduced by 44%.

Keywords: Inspection Interval · Condition-based Maintenance · Two-component series system · Non-homogenous degradation

S.-H. Sheu (Ed.): ICIEA-EU 2024, LNBIP 507, pp. 1–11, 2024.
https://doi.org/10.1007/978-3-031-58113-7_1

1 Introduction

1.1 Machine, Degradation, and Maintenance

Machines are important components of a production system. Through time, these machines degrade, i.e., Degradation has been defined as physical damage or flaws that accumulate during use [1]. Machines in manufacturing systems degrade and eventually fail due to many factors. The factors include but are not limited to (a) natural wear and tear due to continuous usage and (b) damage resulting from mishandling or incorrect usage [2]. Organizations perform maintenance to keep machines in good condition or to restore machines to better condition. Machines in good condition are characterized as those machines that are either new or have negligible flaws, i.e., flaws that do not affect the performance of the machine [2]. Studies such as those by Liao in 2009 and Khatab in 2018 have stated that when the degradation of a machine worsens, the cost to produce using that machine increases [3, 4] This is because of higher reject rates, higher power consumption, and other related factors due to machine degradation.

1.2 Maintenance and General Types

Maintenance programs commonly aim to minimize the total cost of interacting, usually conflicting, costs of maintenance such as preventive maintenance and breakdown maintenance. Park et al. in 2016 defined three general types of maintenance as (a) corrective maintenance, (b) preventive maintenance, and (c) condition-based maintenance [5]. Corrective maintenance performs maintenance activities after every breakdown to restore the machine to good condition. Preventive maintenance performs maintenance activities based on a pre-determined schedule to prevent anticipated breakdown. Lastly, condition-based maintenance performs maintenance activities or does not perform maintenance activities, depending on the findings or data collected during inspection. The data collected are also known as Condition Monitoring or CM data. Examples of CM data as explored in previous studies are (1) degradation in a gearbox [6], oil purity data [7], and vibration data [8]. A usual challenge is being able to identify the correct parameters that will relate to the failure or breakdown of the machine. In the studies above, technical experts such as manufacturers of the components identified the CM data that must be monitored regularly. The correct CM data must be identified for the condition-based maintenance program to be successful. Additionally, condition-based maintenance programs are the most applicable and cost-efficient in an industrial or manufacturing setting, given that the correct parameters are being monitored regularly [9]. As such, determining optimal condition-based maintenance policies is of high relevance to manufacturing systems and because of this, it is the interest and focus of this study.

1.3 Types of Condition-Based Maintenance

Condition-based maintenance, as implemented in practice, has two major types:

(a) *No manual intervention during inspection/collection of data*

- involves collection of CM data continuously and can be done with minimal to no downtime during collection of CM data

(b) *With manual intervention and downtime during inspection and collection of data*

- involves the collection of CM data during inspection intervals

When planning the amount of maintenance, the lowest cost is usually achieved through balancing the lack of maintenance (dealing with breakdown) and excessive maintenance (performing unnecessary activities to prevent breakdown) such as by disassembling a machine when it is not needed or replacing parts that are still working properly. Excessive preventive maintenance can eliminate machine downtime and machine failure; however, this results in a significantly higher cost [10].

Condition-based maintenance (CBM) has been widely used in manufacturing like in semiconductor production facilities. As mentioned above, CBM determines machine degradation by collecting condition monitoring (CM) data through inspections. Specific actual CM data collected by studies and applications include vibration data that was correlated with bearing damage [8] and oil data that was used to estimate the condition of the machine [7]. In 2012, a comparative study on preventive maintenance and condition-based maintenance concluded that the applicability of CBM for many companies is more attainable and cost-efficient, given that the correct parameters are determined for the CBM.

In some cases, the CBM may be simplified into a two-component system based on its critical parts. Once the system is simplified into two components, CBM will be feasible and beneficial because continuous monitoring of all the components is, in general, difficult in a manufacturing setting [11]. For example, in 2018, a study analyzed a wind turbine system's policy to include the two most expensive parts only: (a) the gearbox and (b) the inverter [12]. Also, in 2015, another study predicted the useful life of a slurry pump via the monitoring of two of its most critical parts: the (a) impeller and the (b) motor [3].

1.4 Degradation and Process Rejects

While the primary driver of the cost of breakdown and maintenance is determined by the type of maintenance the company chooses to practice, another significant contributor to the total cost of breakdown and maintenance is process rejects. Process rejects are products created by a system that does not meet the requirements or specifications [4]. Process rejects naturally increase as the machine degrades over time. Furthermore, controlling process rejects becomes more important as the cost to produce a unit of a good product increases [13].

2 The Research Problem

2.1 Problem Statement

The problem that this study tackles is: "How to minimize the total annual condition-based maintenance cost of a two-component series manufacturing system?"

The problem above is based on the high cost of the described system, i.e., maintenance of a two-component series manufacturing system, and on the discussions in the previous sections that explained why condition-based maintenance is of high importance to many companies. Furthermore, as mentioned, if a manufacturing system can be simplified into its two most critical components, CBM may be applied. Moreover, the total cost of maintenance is driven by many factors like the cost of project rejects. In more detail, the cost of CBM can be broken down into (a) inspection, (b) general repair (including replacement), (c) breakdown, and (d) process rejects. Mathematically,

$$\begin{aligned} \textit{Total Annual Cost} = & \textit{Inspection Cost} + \textit{General Repair Cost}\ (\textit{Component}\ 1 + \textit{Component}\ 2) + \\ & \textit{Breakdown Cost}\ (\textit{Component}\ 1 + \textit{Component}\ 2) + \\ & \textit{Process Reject Cost}\ (\textit{Component}\ 1 + \textit{Component}\ 2) \end{aligned} \tag{1}$$

2.2 Objective

The objective of this study is to minimize the overall annual cost of a two-component series system by:

- Determining the frequencies of:

- Inspections (time between inspection intervals)
- General Repairs of each of the two components
- Component Replacements

- Solving for the impact to the cost of the interventions above, in terms of:

- Expected breakdown of components
- Expected process rejects produced by the system

2.3 Hypothesis of This Study

This study hypothesizes that the total cost of a condition-based maintenance policy for a two-component series system can be found by solving jointly for the following:

- Optimizing the inspection interval (inspection frequency)
- Determining the general repair conditions for both components
- Estimating the cost of process rejects and breakdown at every inspection and maintenance done on the system

Performing the above will give a more realistic cost structure for maintenance as encountered in real-life manufacturing systems.

3 Modeling and Formulation

The system is modeled as follows. Please see Fig. 1 for a more detailed illustration.

- Two-component Series System, each component as C1 and C2 respectively
- The machine cannot produce any products when one of the components is not working
- As products are manufactured, process rejects are observed from C1 and C2 separately
- Each component is deteriorating over time due to continuous usage
- Degradation data for each component is inspected at regular intervals
- The current cumulative degradation of each component contributes to the following cost factors:

– Probability of breakdown
– Process rejects from each component

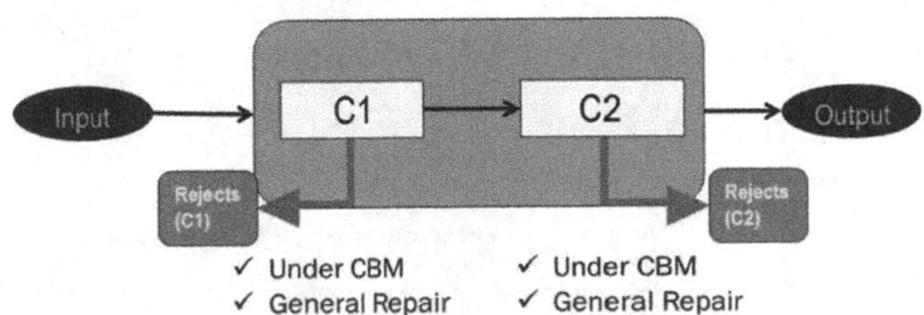

Fig. 1. Illustration of the Two-Component Series System

3.1 Top-Level Representation

In this section, the operating states, maintenance states, and breakdown states are illustrated. Please see Fig. 2.

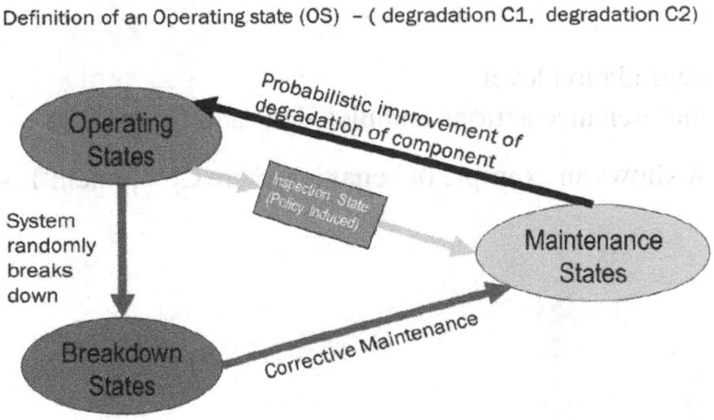

Definition of an Operating state (OS) – (degradation C1, degradation C2)

Operating States
Probabilistic improvement of degradation of component
Inspection State (Policy Induced)
System randomly breaks down
Maintenance States
Breakdown States
Corrective Maintenance

Fig. 2. The Operating States, Maintenance States, and Breakdown States

At any time, the system may move from any operating state to a breakdown. Likewise, from any operating state, the state can move to a maintenance state, i.e., during inspection. Lastly, upon breakdown, the machine component is immediately replaced.

3.2 Repair

Here, the impact of repair on the degradation of a component is discussed. The current degradation of each component is conditional on the following:

1. Previous degradation level
2. Previous maintenance actions completed on each component

Figure 3 below shows two examples of behavior when Component 2 is being repaired.

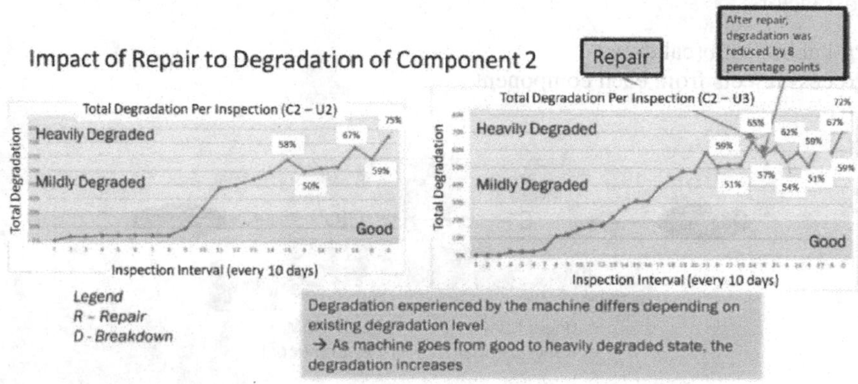

Fig. 3. Two Examples of Behaviors of Component 2 Degradation

3.3 Replace

Instead of repair, severely degraded components are replaced. In this section, the "replace" action is discussed. The current degradation of each component is conditional to the following:

1. The previous degradation level
2. The previous maintenance actions completed in each component

Figure 4 below shows an example of behavior where Component 1 is being replaced.

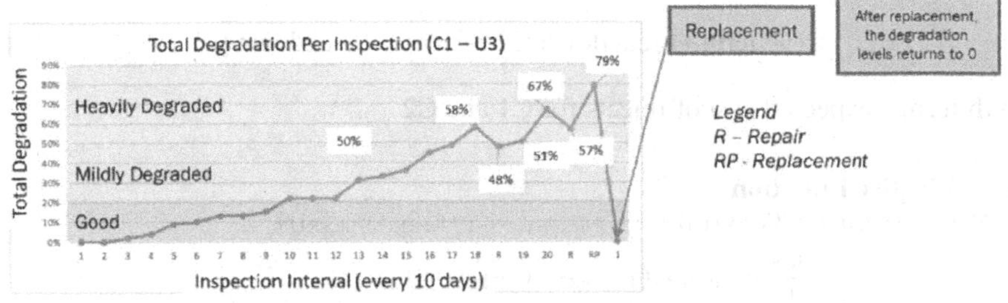

Fig. 4. Example of Behavior of Component 1 Replacement

3.4 Breakdown

In modeling breakdown, the current breakdown probability of each component is conditional to the previous degradation level. Table 1 shows a sample of breakdown occurrences related to the degradation of the component.

Table 1. Sample Breakdown Frequencies

% Degraded Upon Breakdown	C1 Breakdown Frequency	C2 Breakdown Frequency
0–40%	0	2
40%–50%	2	0
50%–70%	5	3
70% and above	4	5
Total Breakdowns	11	10

3.5 Cost Function

Given the foregoing discussions, the cost function is hereby written as:

$$\frac{365}{r} * \pi_{r,(DN)} * (C_I + C_{LP,I}) + \frac{365}{r} * \pi_{r,(PR1)} * (C_I + C_{PR1} + C_{LP,PR1}) + \frac{365}{r} * \pi_{r,(PR2)} * (C_I + C_{PR2} + C_{LP,PR2}) + \frac{365}{r} * \pi_{r,(PT1)} * (C_I + C_{PT1} + C_{LP,PT1}) + \frac{365}{r} * \pi_{r,(PT2)} * (C_I + C_{PT2} + C_{LP,PT2}) + \frac{365}{r} * \pi_{r,(OH)} * (C_I + C_{\mathrm{OH}} + C_{LP,OH}) + \lambda_{BD1} * \pi_{r,BD1} * (C_{CR1} + C_{LP,CR1}) + \lambda_{BD2} * \pi_{r,BD2} * (C_{CR2} + C_{LP,CR2}) + \sum_k \sum_i (\tau_k * RJ_{i,k} * C_{RJ,i}) \quad (2)$$

where:

1) The first 6 terms pertain to the total expected cost from executing each action as a proportion of the total inspections done (at annual number of inspections), where:

1

1^{st}term – inspect and do nothing 4^{th}term – inspect and replace C1
2^{nd}term – inspect and repair C1 5^{th}term – inspect and replace C2
3^{rd}term – inspect and repair C2 6^{th}term – inspect and overhaul

2) Expected cost of breakdown of C1 and C2

$$7^{th}\text{term}-\text{breakdown C1 }8^{th}\text{term}-\text{breakdown C2}$$

3) 9th term – expected cost of rejects for C1 and C2

3.6 Objective Function

$$\begin{aligned}\text{Min}: &\frac{365}{r}*\pi_{r,(DN)}*(C_I+C_{LP,I})+\frac{365}{r}*\pi_{r,(PR1)}*(C_I+C_{PR1}+C_{LP,PR1})+\\&\frac{365}{r}*\pi_{r,(PR2)}*(C_I+C_{PR2}+C_{LP,PR2})+\\&\frac{365}{r}*\pi_{r,(PT1)}*(C_I+C_{PT1}+C_{LP,PT1})+\frac{365}{r}*\pi_{r,(PT2)}*(C_I+C_{PT2}+C_{LP,PT2})+\\&\frac{365}{r}*\pi_{r,(OH)}*(C_I+C_{OH}+C_{LP,OH})+\lambda_{BD1}*\pi_{r,BD1}*(C_{CR1}+C_{LP,CR1})+\lambda_{BD2}*\\&\pi_{r,BD2}*(C_{CR2}+C_{LP,CR2})+\sum_k\sum_i(\tau_k*RJ_{i,k}*C_{RJ,i})\end{aligned} \tag{3}$$

s.t.:

Constraint 1 $\sum_{m\in M}\pi_{r,(m)}=1$
Constraint 2 $\sum_{k\in K}\tau_k=365$
Constraints 3 and 4 $\pi_{r,k(m)},\pi_{r,BD1}\geq 0;\ \pi_{r,k(m)},\pi_{r,BD1}\leq 1$
Constraint 5 $r>0$

where:

- *m: is an element of the defined action space M*
- *The first constraint refers to the total probability that the system is inspected*
- *The second constraint refers to the total sojourn time at each operating state must equal one year (365 days)*
- *The third and fourth constraint states that probability values should be between 0 to 1 only*

The fifth constraint states that the inspection interval r must be greater than 0

$$\begin{aligned}\text{Min}: &\frac{365}{r}*\pi_{r,(DN)}*(C_I+C_{LP,I})+\frac{365}{r}*\pi_{r,(PR1)}*(C_{PR1}+C_{LP,PR1})+\frac{365}{r}*\pi_{r,(PR2)}*(C_{PR2}+C_{LP,PR2})+\\&\frac{365}{r}*\pi_{r,(PT1)}*(C_{PT1}+C_{LP,PT1})+\frac{365}{r}*\pi_{r,(PT2)}*(C_{PT2}+C_{LP,PT2})+\\&\frac{365}{r}*\pi_{r,(OH)}*(C_{PT2}+C_{LP,OH})+\lambda_{BD1}*\pi_{r,BD1}*(C_{CR1}+C_{LP,CR1})+\lambda_{BD2}*\\&\pi_{r,BD2}*(C_{CR2}+C_{LP,CR2})+\sum_k\sum_i(\tau_k*RJ_{i,k}*C_{RJ,i})\end{aligned} \tag{4}$$

s.t.:

Constraint 1 $\sum_{m\in M}\pi_{r,(m)}=1$
Constraint 2 $\sum_{k\in K}\tau_k=365$
Constraints 3 and 4 $\pi_{r,k(m)},\pi_{r,BD1}\geq 0;\ \pi_{r,k(m)},\pi_{r,BD1}\leq 1$
Constraint 5 $r>0$

Decision Variables (to be jointly optimized)

1). *Inspection interval r*
2). $\pi_{r,(m)}$ *- proportion of the total inspections that will be assigned to each maintenance action m*

4 Proposed Methodology for Implementation in Manufacturing Systems

Given all of the above discussions, especially the modeling of the system with the objective function, this study proposes to implement CBM using the following methodology as follows:

A. Model the Two-Component System

I. Set up the Operating States
II. Set up the Breakdown States
III. Define the System Costs

B. Establish the Decision Variables

I. Establish the Inspection State
II. Establish the Maintenance States in relation to the CM data

C. Perform the Exhaustive Monte Carlo approach to the Optimal Decision Variables
D. Use the Identified Optimal Decision Variables from C above

This general methodology is detailed in Fig. 5 below.

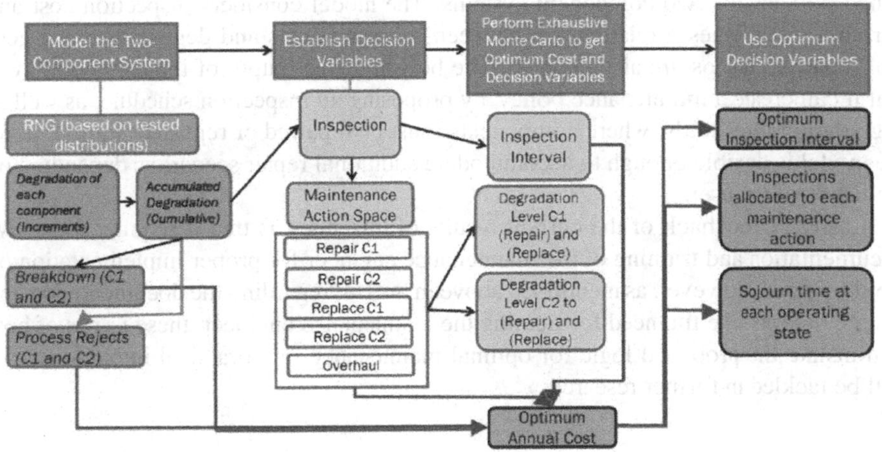

Fig. 5. Proposed Methodology for Implementing in Manufacturing Systems

Important Implementation Issues for the Above Methodology:

The above-proposed methodology may result in a concern about the amount of documentation needed such as for various information about degradation, reject, inspection, and schedules of inspections. Training the maintenance engineer on the implementation of the proposed methodology may also be perceived as a challenge, as it will require specialized knowledge of maintenance optimization. However, this study aims to tackle the logic for the optimal maintenance system. A subsequent study will aim for the translation of the proposed methodology for practical use, which may either be the translation of this methodology into easy-to-implement and easy-to-use reminders and templates or into a fully automated alert and guide system to signal the need for maintenance and guide in the recommended maintenance activities.

5 Case Study

A case study was performed on a local manufacturing company of Fast-Moving Consumer Goods in the Philippines. With almost 4 years of data, the proposed methodology of this study optimized the repairs and replacements of the two components of the company's cutter (C1)-grinder (C2) assembly. By doing so, simulations for long-term results showed an average projected savings of 44% in total annual maintenance costs. Breakdowns and process rejects were effectively jointly decreased, and stay times in the degraded states were also minimized. Another paper will discuss thoroughly and in more detail the case study performed.

6 Discussions

The proposed methodology and objective function optimize condition-based maintenance policies for two-component systems. The model considers inspection cost and duration, establishes a relationship between process rejects and degradation and general repair scenarios are also covered. The benefit of the output of the methodology is that it can create a maintenance policy, by proposing an inspection schedule, as well as degradation thresholds, where components must be repaired or replaced. Furthermore, the model is flexible enough to accommodate additional repair scenarios, depending on the needs of the user.

Lastly, a drawback of the current results of this study is that it requires extensive documentation and training of the maintenance engineer for proper implementation on the shop floor. However, as mentioned above in Sect. 4, regarding the documentation and other concerns like the need for training the maintenance engineer, these issues of how to translate the proposed logic for optimal maintenance into practical implementation will be tackled in further research.

7 Conclusion

A condition-based maintenance model for two components subject to non-homogenous degradation processes that integrate frequency of inspections, process rejects, and general repair scenarios resulted in minimizing the overall total annual cost. An exhaustive Monte Carlo simulation on the proposed model was used to solve for the optimum policy and results showed an average projected savings of 44% in total annual maintenance costs.

References

1. Xu, R.-Z., Xie, L., Zhang, M.-C., Li, C.-X.: Machine degradation analysis using fuzzy CMAC neural network approach. Int. J. Adv. Manuf. Technol. **36**(7–8), 765–772 (2008). https://doi.org/10.1007/s00170-006-0887-6
2. Naderkhani ZG, F., Makis, V.: Optimal condition-based maintenance policy for a partially observable system with two sampling intervals. Int. J. Adv. Manuf. Technol. **78**(5–8), 795–805 (2015). https://doi.org/10.1007/s00170-014-6651-4
3. Liao, G.-L., Chen, Y.H., Sheu, S.-H.: Optimal economic production quantity policy for imperfect process with imperfect repair and maintenance," Eur. J. Oper. Res. **195**(2), 348–357 (2009). https://doi.org/10.1016/j.ejor.2008.01.004
4. Khatab, A.: Maintenance optimization in failure-prone systems under imperfect preventive maintenance. J. Intell. Manuf. **29**(3), 707–717 (2018). https://doi.org/10.1007/s10845-018-1390-2
5. Park, C., Moon, D., Do, N., Bae, S.M.: A predictive maintenance approach based on real-time internal parameter monitoring. Int. J. Adv. Manuf. Technol. **85**(1–4), 623–632 (2016). https://doi.org/10.1007/s00170-015-7981-6
6. Jafari, L., Makis, V.: Joint optimization of lot-sizing and maintenance policy for a partially observable two-unit system. Int. J. Adv. Manuf. Technol. **87**(5–8), 1621–1639 (2016). https://doi.org/10.1007/s00170-016-8556-x
7. Wang, Y., Deng, C., Wu, J., Xiong, Y.: Failure time prediction for mechanical device based on the degradation sequence. J. Intell. Manuf. **26**(6), 1181–1199 (2015). https://doi.org/10.1007/s10845-013-0849-4
8. Lim, R., Mba, D.: Fault detection and remaining useful life estimation using switching Kalman filters. In: Tse, P.W., Mathew, J., Wong, K., Lam, R., Ko, C.N. (eds.) Engineering Asset Management - Systems, Professional Practices and Certification. LNME, pp. 53–64. Springer, Cham (2015). https://doi.org/10.1007/978-3-319-09507-3_6
9. Ahmad, R., Kamaruddin, S.: An overview of time-based and condition-based maintenance in industrial application. Comput. Ind. Eng. **63**(1), 135–149 (2012). https://doi.org/10.1016/j.cie.2012.02.002
10. Zheng, X., Fard, N.: A maintenance policy for repairable systems based on opportunistic failure-rate tolerance. Microelectron. Reliab. **32**(4), 237–244 (1992). https://doi.org/10.1016/0026-2714(92)90521-L
11. Tse, P.W., Shen, C.: Remaining useful life estimation of slurry pumps using the health status probability estimation provided by support vector machine. In: Tse, P.W., Mathew, J., Wong, K., Lam, R., Ko, C.N. (eds.) Engineering Asset Management - Systems, Professional Practices and Certification. LNME, pp. 87–98. Springer, Cham (2015). https://doi.org/10.1007/978-3-319-09507-3_9
12. Nourelfath, M., Nahas, N., Ben-Daya, M.: Integrated preventive maintenance and production decisions for imperfect processes. Reliab. Eng. Syst. Saf. **148**, 21–31 (2016). https://doi.org/10.1016/j.ress.2015.11.015
13. Do, P., Assaf, R., Scarf, P., Iung, B.: Modelling and application of condition-based maintenance for a two-component system with stochastic and economic dependencies. Reliab. Eng. Syst. Saf. **182**, 86–97 (2019). https://doi.org/10.1016/j.ress.2018.10.007

Determination of Skilled Worker Requirements for Maintenance Departments Under Stochastic Failure Mode Conditions

Sahin Furkan Sahiner(✉) and Onur Golbasi

Middle East Technical University, Üniversiteler Mahallesi, Dumlupınar Bulvarı No:1, 06800 Çankaya, Ankara, Turkey
{fsahiner,golbasi}@metu.edu.tr

Abstract. Maintenance within the mining industry is a global challenge, demanding innovative solutions due to its multifaceted nature. The complexity, size, competition, escalating costs, and the need for a proficient workforce present formidable obstacles. This study focuses on optimizing technically skilled worker allocation in the mining industry's maintenance department, developing a continuous-event simulation model. The ever-increasing intricacy of mining equipment necessitates heightened reliability to ensure efficient, continuous production. Adequate maintenance is crucial in this context. The research problem revolves around the critical role of technically skilled workers during maintenance operations, where their absence can lead to delays and operational inefficiencies. Skilled maintenance technicians are indispensable, capable of accurately diagnosing equipment issues, reducing costly breakdowns, and minimizing downtime through preventive maintenance. They also expedite the repair process when required. To address these challenges, our study introduces a continuous event simulation algorithm designed to minimize costs. This algorithm takes into account factors such as production losses and maintenance workforce expenses while maximizing equipment utilization. By doing so, the research contributes to the field by emphasizing the importance of a skilled workforce in mining maintenance, ensuring equipment longevity, performance, and safety. The contribution of this study lies in its practical application of advanced algorithms to optimize technically skilled worker allocation, mitigating operational challenges and highlighting the crucial role of skilled maintenance management in the evolving landscape of mining equipment.

Keywords: Optimization · Event Simulation · Maintenance · Workforce

1 Introduction

Maintenance is a global undertaking that commands an annual expenditure reaching billions of dollars. This challenge has persisted since the dawn of the Industrial Revolution, and it remains multifaceted, demanding innovative approaches to overcome its complexities. Despite substantial progress in enhancing the effectiveness of field maintenance practices, it remains an imposing task due to various factors, such as equipment

S.-H. Sheu (Ed.): ICIEA-EU 2024, LNBIP 507, pp. 12–22, 2024.
https://doi.org/10.1007/978-3-031-58113-7_2

complexity, sheer size, intense competition, escalating costs, safety concerns, and, most importantly, improper maintenance workforce. Simultaneously, the mining industry, deeply rooted in human civilization for centuries, has played a pivotal role in supplying the essential raw materials needed for industry and infrastructure. Presently, a significant workforce is engaged in the global mining industry, with the United States, for instance, employing approximately 675,000 [1] individuals in the natural resources and mining field. Throughout its evolution, mining operations have undergone transformative changes, with the complexity and sophistication of the equipment used standing out as a paramount shift. As mining equipment becomes progressively more intricate, it also experiences a steep and consistent cost increase. This upward cost trajectory presents a formidable challenge for mining companies, particularly in the context of standby units that have become increasingly cost-ineffective. In response to these challenges, mining companies are prompted to demand heightened equipment reliability. Reliability stands as a crucial performance metric, reflecting the overall condition of equipment and ensuring that it consistently delivers satisfactory performance over the desired operational lifespan, adhering to specified requirements. This simply means equipment management is the cornerstone of efficiency and cost-effectiveness. The reliability and lifespan of equipment directly impact the productivity and budget of operations. An essential aspect of this management is maintenance, and the effectiveness of maintenance efforts is contingent on various factors. Among these factors, the presence and proficiency of skilled workers stand out as a critical determinant in the success of maintenance endeavors. An adequate number of skilled workers plays a pivotal role in sustaining equipment and reducing the associated maintenance costs.

In the maintenance and repair time realm, it is valuable to deconstruct and comprehend the various constituent elements at play. Figure 1 is an illustrative aid in elucidating the typical composition of maintenance repair time. This depiction underscores that the actual time dedicated to executing a maintenance repair task often constitutes only a fraction of the overall downtime. The totality of downtime encompasses several contributing factors. As depicted, "supply delay" and "maintenance delay" represent substantial portions of the total downtime. During the supply delay phase, spare parts are procured for the equipment, entailing supply chain inefficiencies, logistics considerations, and procurement lead times.

Upon the arrival of equipment spare parts, the groundwork for maintenance or replacement work is nearly set. Nevertheless, this juncture places substantial reliance on an adequate number of technically skilled workers to orchestrate and initiate the maintenance process. During this phase, potential maintenance delays can manifest, comprising administrative lead times and hindrances stemming from a shortage of maintenance human resources. These multifaceted dynamics underscore the intricate nature of maintenance repair time and its broader implications on operational efficiency.

From the mining industry perspective, the mining industry relies heavily on a diverse array of machinery and equipment for extracting valuable minerals and resources from both surface and underground environments. The maintenance of these assets is not merely a routine task; it is a critical factor in ensuring their longevity, optimal performance, continuous production, and the overall success of mining operations. Maintenance departments are core components for sustaining production in the mining industry.

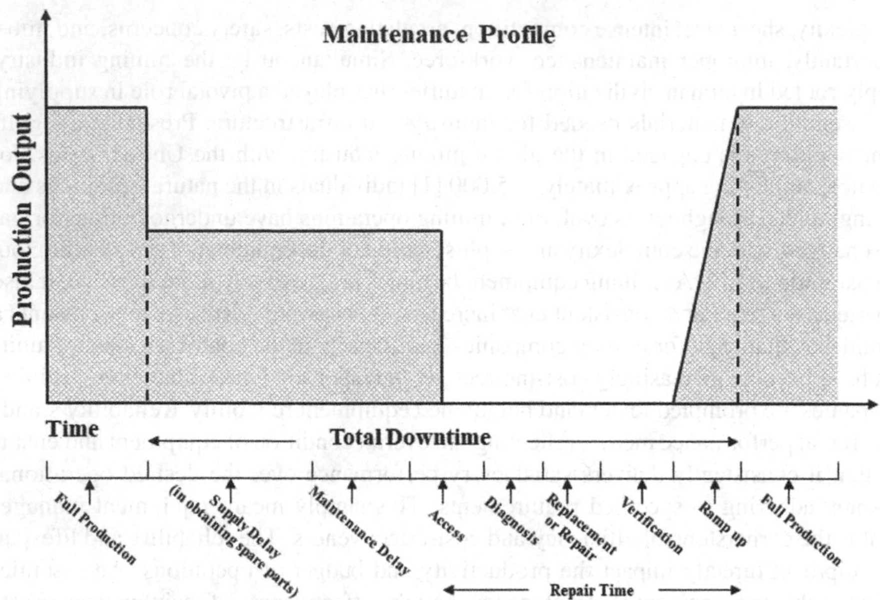

Fig. 1. Maintenance Repair Time [2].

This is because hundreds of equipment with different operational functionalities must be maintained above the administrative thresholds while exposed to various failure modes and profiles. In Fig. 2, it is shown that how maintenance and production actually have two intertwined parts.

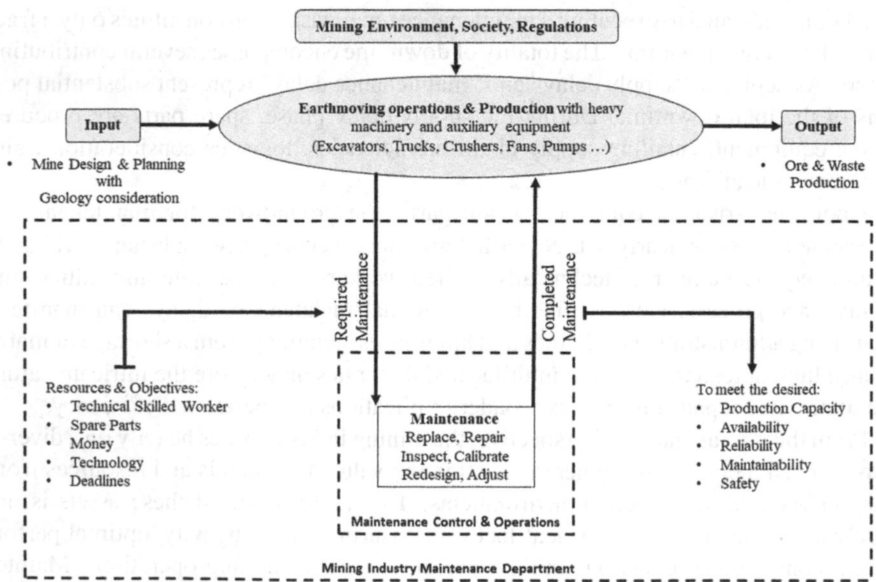

Fig. 2. Interconnected Spheres of Maintenance and Production in Mining Operations.

Large-scale machinery such as excavators, haul trucks, bulldozers, and conveyors play a pivotal role in surface mining. Regular maintenance is essential to ensure their continued operation. Maintenance entails inspections, servicing, and the replacement of worn parts to maintain safety and prevent costly breakdowns. Skilled maintenance technicians are vital in diagnosing issues, conducting repairs, and performing preventive maintenance. Their expertise not only keeps the equipment running efficiently but also minimizes downtime, thereby contributing to increased productivity. In underground mining operations, specialized equipment is required to navigate the unique challenges presented by confined spaces and harsh conditions. Machinery such as underground drills, load-haul-dump (LHD) machines, and continuous miners must undergo meticulous maintenance to ensure safety and prevent hazardous situations. In underground mining, equipment breakdowns can lead to dangerous scenarios and mine closures. A skilled workforce is pivotal in conducting timely inspections, servicing, and overhauls to guarantee that the equipment remains reliable and safe. That is, the maintenance of mining equipment, whether on the surface or in underground mines, is a linchpin of the industry's operational success. It safeguards not only the efficiency and longevity of machinery but also the safety of the workforce. A highly skilled technical workforce is indispensable for this maintenance effort. Their expertise in diagnosing and rectifying issues, combined with regular inspections and preventive maintenance, ensures that mining equipment remains operational and contributes to the industry's productivity and, above all, the well-being of its personnel. The harmonious interplay between maintenance and a skilled workforce is essential for the continued prosperity and safety of the mining industry.

In this context, enough skilled workers play a pivotal role in maintenance by leveraging their expertise in accurately and swiftly diagnosing equipment issues, thereby preventing costly breakdowns and minimizing production downtime. Their adeptness at executing preventive maintenance tasks extends the operational life of the equipment and reduces the frequency of repairs, enhancing overall equipment reliability. Furthermore, in cases where equipment requires repair, skilled workers efficiently expedite the process, minimizing downtime and mitigating associated production losses. Their capacity to develop and implement optimized maintenance schedules ensures that equipment is serviced at the correct intervals, aligning with manufacturer recommendations and industry best practices, thus reducing the likelihood of unexpected failures and promoting smooth, uninterrupted operations.

In this sense, the current study aims to develop a continuous event simulation algorithm minimizing the cost items, including production losses and maintenance workforce expenses, while maximizing the equipment utilization by considering the proper number of skilled maintenance workers for the mining industry.

2 Literature Review

Maintenance management strives to improve production system availability and decrease the frequency of failures, all while minimizing operational costs and losses due to failures. Two widely recognized maintenance categories are corrective and preventive maintenance. Preventive maintenance is conducted while the system is still operational, whereas corrective maintenance is implemented in response to a system failure. Whether in industry or transportation, maintenance managers seek to optimize preventive maintenance planning to simultaneously reduce the occurrence of corrective maintenance and minimize overall maintenance costs [3].

The mining industry relies heavily on effective equipment maintenance to ensure optimal operational efficiency and safety. This literature review provides an overview of various studies on maintenance practices within the mining sector. These studies encompass a wide range of methodologies and approaches, offering valuable insights into maintenance decision-making, cost optimization, reliability enhancement, and the analysis of maintenance-related data in mining operations.

In a study by Barberá et al. [4], the authors introduced the GAMM (Graphical Analysis for Maintenance Management) method, which facilitates maintenance decision-making through graphical data analysis. This method was applied to evaluate deficiencies and propose improvements in the maintenance of slurry pumps in a Chilean mining plant. Golbasi and Turan [5] presented a maintenance policy optimization approach for establishing the most effective maintenance work packages by considering the equipment's uptime and downtime characteristics. Ali and Reza [6] developed a method using statistical models, including univariate exponential regression (UER) and multivariate linear regression (MLR), to determine the overhaul and maintenance cost of loading equipment in surface mining. Their model considered various equipment parameters, providing a practical tool for cost-related decision-making. Morad et al. [7] investigated the maintenance policy performance of operating trucks in Sungun Copper Mine, focusing on minimizing failure downtimes. Their study demonstrated the potential for extended operational hours through failure and maintenance profile analysis, system reliability assessment, and Monte Carlo Simulation. Kovacevic et al. [8] introduced a two-step method to analyze factors influencing human errors during mining machine maintenance. Cause-effect analysis and group fuzzy analytic hierarchy process were employed to reveal crucial aspects related to work instructions, training, work experience, and equipment specifications. Nikulin et al. [9] presented a computer-aided application for evaluating operational and maintenance strategies, constructing various scenarios to correlate system reliability, maintainability, and operational behavior. Gölbaşı and Demirel [10] developed a "time-counter" simulation algorithm to minimize maintenance costs and optimize inspection intervals for mining machines. The model was applied to draglines in a coal mine, resulting in notable cost reductions. Jonsson et al. [11] discussed the analysis of digitalized condition-based maintenance data from machinery in an iron ore mine in Sweden, emphasizing the significance of digital representations in work practices. Angeles and Kumral [12] proposed a maintenance management approach for the mining industry, focusing on enhancing equipment availability, reliability, and proactive failure prevention. Their methodology utilized failure data from a mining truck fleet in a Canadian open-pit mine to determine optimal inspection intervals.

These studies collectively contribute to the knowledge of mining equipment maintenance, offering practical tools, methodologies, and data-driven approaches to improve maintenance decision-making, optimize costs, and enhance equipment reliability in the mining sector. In light of the preceding research, our study identifies a notable gap in the existing literature. Specifically, it addresses the complex task of determining the optimal number of workers responsible for diverse skill sets and managing various failure types characterized by random lifetimes and multiple failure modes. Through the application of event simulation, our work aims to bridge this gap and contribute to future studies in this field.

3 Methodology

As depicted in Fig. 3, the comprehensive research methodology comprises three key phases. In the initial step, the system under investigation, including its machinery or equipment, is precisely identified, with a meticulous assessment of its components and associated failure modes. The failure modes are systematically categorized, while lifetime and repair time data are rigorously processed [10], integrating insights from domain experts and relevant documentation. System constraints are further delineated, and the processed failure dataset is incorporated into the algorithm. In the subsequent step, a specialized simulation algorithm tailored to address failure-specific scenarios is meticulously developed using BlockSim® software, culminating in the formulation of a technical skilled workforce allocation strategy. In the final phase, the model is applied to a practical case, wherein random operating and repair times are closely monitored, facilitating the analysis of production losses and the overall system cost. The optimization of technical skilled worker requirements is achieved by meticulously examining various scenarios, and the resulting strategy is comprehensively documented for further evaluation and implementation.

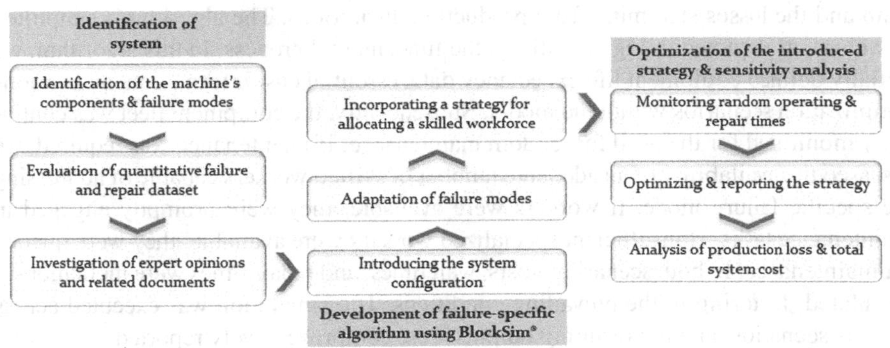

Fig. 3. Research Methodology Flowchart

3.1 Model Development

The developed algorithm is a continuous, stochastic, and dynamic simulation model for system maintenance. It continuously monitors the system status with predefined intervals, capturing variable values and making decisions based on critical variables randomly assigned from distribution functions. The dynamic structure emphasizes the time-based evaluation of system changes. Skilled worker requirement for maintenance activities is determined based on real-time worker information. The algorithm accounts for a predetermined maintenance personnel number for each failure mode, allowing for the potential allocation of workers. The simulation starts at time zero and progresses in time increments until reaching a target simulation time. Failure mode occurrence times are randomly assigned, and the algorithm considers various functional dependencies between failure modes, including series and parallel configurations. Maintenance activities are initiated when failure modes surpass their lifetime finish points. Skilled worker unavailability leads to interruptions in production. The algorithm continuously updates maintenance finish points based on maintenance activities and their modifications. New lifetime finish points are assigned when the maintenance is completed. The algorithm captures information on maintenance personnel availability, maintenance finish points, and equipment conditions. Cumulative values for equipment availabilities, total downtime, downtime due to worker unavailability, and cost metrics are stored for each piece of equipment. In summary, the algorithm provides a detailed continuous event model for system maintenance, considering stochastic factors, skilled worker dynamics, and dynamic system changes over time, with a focus on optimizing maintenance personnel allocations and minimizing cost items.

In this study, the primary objective was to optimize the allocation of technically skilled workers within the maintenance department, responsible for the upkeep of a diverse fleet of mining equipment susceptible to various failure modes, utilizing continuous event simulation. These skilled workers possess distinct competencies tailored to address specific failure modes. We aimed to ascertain the ideal number of workers across different skill groups, maximizing operational efficiency while minimizing costs. This consideration encompassed the expenses associated with the maintenance team and the losses stemming from production disruptions. The algorithm's simplified flowchart, as depicted in Fig. 4, outlines the fundamental process. In this algorithm, we initially defined equipment life expectancy data, essential cost information, and various optimization scenarios within the model. Subsequently, the equipment fleet was continually monitored for the need for random maintenance. If maintenance was required, we assessed the availability of an adequate number of skilled workers capable of addressing the specific failure mode. If workers were available, they were promptly engaged in maintenance tasks. If insufficient specialized workers were available, they were queued for maintenance. In both scenarios, costs, wait times, and repair times were meticulously calculated, factoring in the prevailing conditions. The simulation was executed across various scenarios, and the resulting outputs were comprehensively reported.

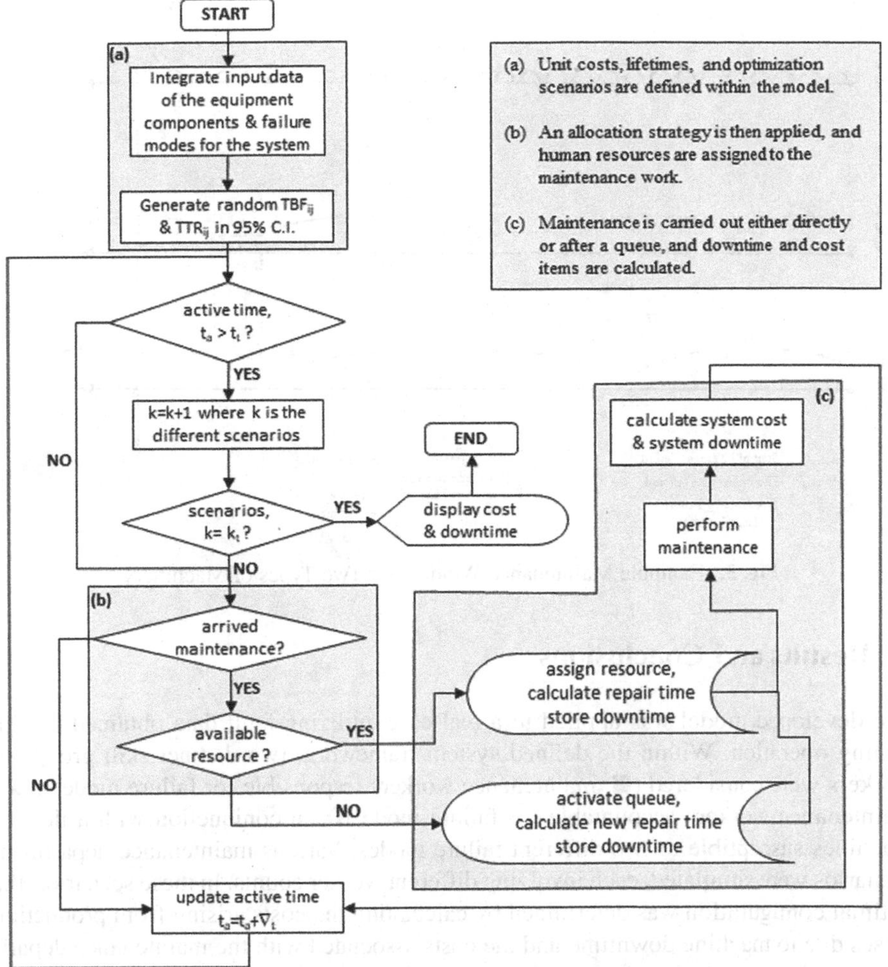

Fig. 4. Flowchart of the Developed Algorithm

An illustrative example depicting the logical flow and decision-making processes involved in maintenance tasks and crew allocations is presented in Fig. 5. The system exemplified in Fig. 5 encompasses two distinct failure modes occurring on two different pieces of equipment, all managed by a common maintenance department. While the maintenance department may comprise multiple divisions, this specific example assumes that a particular part of the department handles all these types of failures. Other divisions, if present, are not considered suitable for intervening in repair works. As demonstrated in the example, inadequate worker availability can lead to extended downtime, particularly in cases involving multiple simultaneous faults. This, in turn, results in production losses and an upsurge in total costs. To mitigate such situations, the idea of hiring a larger workforce may arise. However, this approach entails an increase in maintenance department expenses. In summary, a delicate balance must be struck between these contrasting considerations.

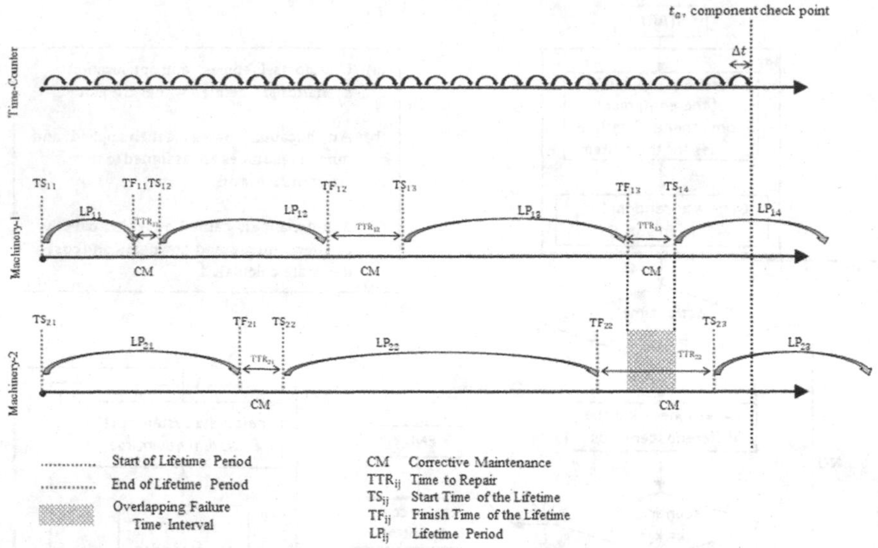

Fig. 5. Example Maintenance Window for Two Types of Machinery

4 Results and Conclusions

The developed model was applied to a real case, utilizing fault data obtained from a mining operation. Within the defined system framework, two distinct skill groups of workers were considered (■: maintenance workers responsible for failure mode A, ▲: maintenance workers accountable for failure mode B), in conjunction with a fleet of machines susceptible to two different failure modes. Various maintenance department scenarios were simulated, each involving different worker counts. In these scenarios, the optimal configuration was determined by calculating the costs arising from production losses due to machine downtime and the costs associated with the maintenance department workforce. The results have been visually represented in Fig. 6. Consequently, within the defined system boundaries, the scenario that yielded the lowest total system cost involved ▲ three workers with technical skills and ■ five workers with specialized skills.

In summary, maintenance departments manage hundreds of equipment units susceptible to various failure modes, and the efficacy of skilled workers in diagnosing issues, conducting preventive maintenance, and expediting repairs are paramount. Their expertise minimizes downtime, prevents costly breakdowns, and enhances overall reliability while reducing maintenance costs. The mining process, whether on the surface or underground, demands a diverse machine fleet for tasks like material loading, hauling, ground drilling, and supporting. Each machine faces multiple failure modes with varying frequencies and severity levels, making the optimal configuration of maintenance personnel crucial. On this basis, this study highlights the essential interplay between maintenance and a skilled workforce in the mining sector, emphasizing the importance of skilled maintenance management in ensuring equipment longevity, performance, and safety.

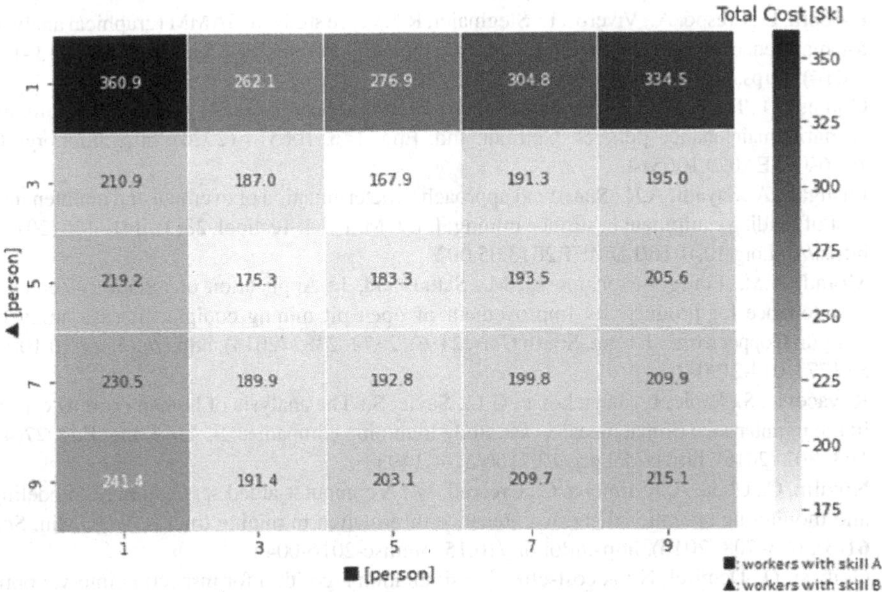

Fig. 6. Simulation Results for Minimizing Total System Costs.

As the industry evolves, technical skills and advanced algorithms will remain central to its enduring role in the global supply chain.

This study presents a versatile continuous simulation model to help the decision-makers in terms of the determination of skilled worker requirements under stochastic failure mode conditions in the mining industry. To enrich future research outcomes, several recommendations are proposed. The model could be extended to include preventive maintenance and other maintenance activity types, fostering a more comprehensive representation. Additionally, recognizing the crucial role of spare part inventory in maintenance, the model could be refined by integrating spare parts inventory policies in subsequent studies to gauge their impact on maintenance downtimes. Further improvements may involve introducing additional constraints or variables related to operational and environmental conditions, such as long-term production plans influenced by climate and market demand fluctuations. Lastly, future studies could delve deeper into failure modes, conducting a more comprehensive analysis of equipment life and repair times to enhance competency groups within the same maintenance division.

References

1. Industries at a Glance: Mining (except Oil and Gas): NAICS 212: U.S. Bureau of Labor Statistics. (n.d.). https://www.bls.gov/iag/tgs/iag212.htm. Accessed 15 Oct 2023
2. Deighton, M.G.: Maintenance Management. Fac. Integri. Manage. 87–139 (2016). https://doi.org/10.1016/B978-0-12-801764-7.00005-X
3. Hamzaoui, A.F., Malek, M., Zied, H., Sadok, T.: Optimal production and non-periodic maintenance plan for a degrading serial production system with uncertain demand. Int. J. Syst. Sci.: Oper. Logist. **10**(1) (2023). https://doi.org/10.1080/23302674.2023.2235808

4. Barberá, L., Crespo, A., Viveros, P., Stegmaier, R.: A case study of GAMM (graphical analysis for maintenance management) in the mining industry. Reliab. Eng. Syst. Saf. **121**, 113–120 (2014). https://doi.org/10.1016/j.ress.2013.07.017
5. Golbasi, O., Turan, M.O.: A discrete-event simulation algorithm for the optimization of multi-scenario maintenance policies. Comput. Ind. Eng. **145**, 106514 (2020). https://doi.org/10.1016/J.CIE.2020.106514
6. Lashgari, A., Sayadi, A.R.: Statistical approach to determination of overhaul and maintenance cost of loading equipment in surface mining. Int. J. Min. Sci. Technol. **23**(3), 441–446 (2013). https://doi.org/10.1016/J.IJMST.2013.05.002
7. Morad, A.M., Pourgol-Mohammad, M., Sattarvand, J.: Application of reliability-centered maintenance for productivity improvement of open pit mining equipment: case study of Sungun Copper Mine. J. Cent. South Univ. **21**(6), 2372–2382 (2014). https://doi.org/10.1007/S11771-014-2190-2
8. Kovacevic, S., Papic, L., Janackovic, G.L., Savic, S.: The analysis of human error as causes in the maintenance of machines: a case study in mining companies. S. Afr. J. Ind. Eng. **27**(4), 193–202 (2016). https://doi.org/10.7166/27-4-1493
9. Nikulin, C., Ulloa, A., Carmona, C., Creixell, W.: A computer-aided application for modeling and monitoring operational and maintenance information in mining trucks. Arch. Min. Sci. **61**(3), 695–708 (2016). https://doi.org/10.1515/amsc-2016-0048
10. Gölbaşı, O., Demirel, N.: A cost-effective simulation algorithm for inspection interval optimization: an application to mining equipment. Comput. Ind. Eng. **113**, 525–540 (2017). https://doi.org/10.1016/J.CIE.2017.09.002
11. Jonsson, K., Mathiassen, L., Holmströ, J., Jonsson, K.: Representation and mediation in digitalized work: evidence from maintenance of mining machinery (n.d.). https://doi.org/10.1057/s41265
12. Angeles, E., Kumral, M.: Optimal inspection and preventive maintenance scheduling of mining equipment. J. Fail. Anal. Prev. **20**(4), 1408–1416 (2020). https://doi.org/10.1007/s11668-020-00949-z

Empirical Findings on the Current State of Industrial Production Management Systems in the Context of Increasing Digitalization

Stefan Schmid(✉) and Herwig Winkler

Brandenburg University of Technology Cottbus-Senftenberg, Konrad-Wachsmann-Allee 13, 03046 Cottbus, Germany

Stefan.Schmid@b-tu.de

Abstract. Increasing digitization and automation are currently bringing a revolution to many processes and sectors of the economy. The industry of production in particular is undergoing a fundamental transformation through the implementation of digital and networked system elements. In production control, traditionally based on established methods and experience, advanced technologies such as artificial intelligence are opening up entirely new options. Artificial intelligence's ability to analyze large amounts of data in real time and make instant decisions from it brings fresh perspectives to production control. Until now, there have been only a few surveys that examine the current status as well as future requirements for production control. Our research focuses on the current state of IT systems used to support decision-makers in production control. It also sheds light on the requirements needed to improve these systems, especially in light of advancing technological developments and increasing digitalization. The article is dedicated to the results and provides an overview of future expectations.

Keywords: Production Management System · Production Control · Manufacturing System · Production · Industry 4.0

1 Introduction

Digital transformation has profoundly changed and revolutionized the modern business world in almost all industries [1]. In industrial production, digitalization has made production control in particular a central element that determines the competitiveness of companies [2]. The age of digitalization is bringing about new technologies, processes and approaches that are fundamentally changing the way companies control and optimize their production processes [3].

In recent years, digital technologies have experienced unstoppable expansion and fundamentally changed the way production processes can be controlled [4]. The integration of technologies such as the Internet of Things (IoT) allows the collection of extensive amounts of data directly from the production environment, resulting in real time monitoring and control [3][5]. This not only enables more accurate tracking of production status, but also early detection of bottlenecks and effective management of

S.-H. Sheu (Ed.): ICIEA-EU 2024, LNBIP 507, pp. 23–34, 2024.
https://doi.org/10.1007/978-3-031-58113-7_3

resources. Linking production machinery and equipment with sensors and smart devices creates a comprehensive ecosystem in which physical processes seamlessly merge with digital data streams [6].

Another emerging approach that is gaining importance in the context of production control is the application of artificial intelligence (AI), specifically machine learning. By analyzing large amounts of data, AI models can identify patterns and relationships that are difficult for human experts to understand [7]. This enables more accurate forecasting of changes in demand, optimization of production schedules, and adaptation to changing market conditions. For example, AI can be used to automatically optimize production processes by taking into account parameters such as machine running times and material availability. This leads to increased efficiency and reduction of production downtime [8].

In order to promote the processes of control and decision-making in manufacturing more effectively, innovative approaches are required that expand the limitations of traditional control and optimization methods of production systems to include aspects and potentials of Industrie 4.0. The article presents an excerpt of the study results from production control in the digital age from the perspective of industrial practice [9]. The objective of the study was to explore the current state of industrial systems for production control. It shows how decision-makers are supported, what challenges are existing and how production control can be improved against the backdrop of advanced technologies and increasing digitalization. The results form the basis and prerequisites for future considerations of a smart production management system to be used for both decision support and auto-control.

2 Methodology

In the field of empirical social research, a distinction can be made between qualitative research methods, which are particularly suitable for explorative research concepts, and quantitative research methods, which are particularly suitable for descriptive objectives [10]. Within empirical social research, different methods can be found, e.g. experiment, survey, archival analysis, historiography, and case study research [11]. The appropriate method must be selected and adapted for the corresponding and individual research process, since there is no generally valid and uniform procedure within empirical social research [12].

A descriptive study is suitable for the presented investigation, since hypotheses already exist and the research field is known. With the established research question the requirements from the industrial environment are to be taken up. In this respect, the survey is seen as an effective tool, since it is suitable for current topics and the result is not influenced by the researcher.

In designing the empirical study, we followed the method of Dieckmann [10], which proposes a structured and systematic five-stage process (illustrated in Fig. 1). The process begins with the formulation of a research question and specification of the research problem. The hypotheses guide the entire study and are central to the subsequent investigations. Once the research question and hypotheses have been determined, the survey is planned and prepared. This step forms the basis for data collection and includes the development and structure of the questionnaire. The selection of a suitable sample is the next important step. In our case, the participants operate either directly or indirectly in the production environment and have relevant experience.

Data can be collected using various methods such as surveys, experiments, archive analyses or case studies. This study was conducted as an online survey, as a specific target group could be reached quickly and statistical evaluations can be recorded effectively. Data collection is followed by data evaluation. In this phase, statistical methods are used to evaluate the collected data. The results of the data analysis are then interpreted and discussed in the context of the research questions in order to understand the significance of the results. Finally, reporting is carried out, which includes a detailed presentation of the results, discussions and conclusions.

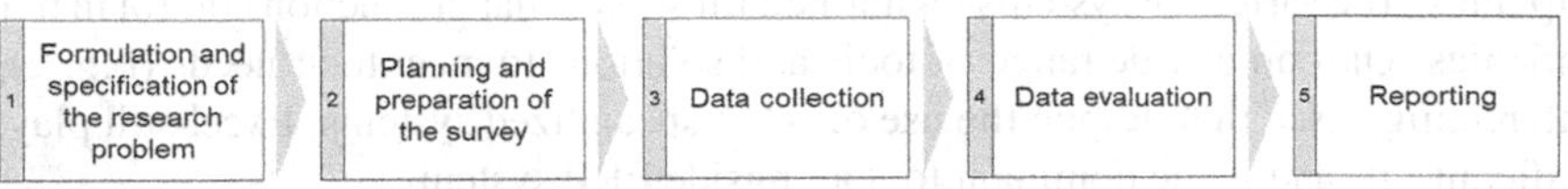

Fig. 1. Procedure of the empirical study [10]

3 Results

The collection of the underlying data was conducted between March and May 2023. A total of 294 individuals participated in the survey, with 125 completing the questionnaire in full. This corresponds to a response rate of 9.36% in relation to the 1336 individuals approached in person.

The vast majority of participants came from the mechanical engineering sector (33.6%), followed by participants from the automotive engineering and automotive supply industry (18.4%). Metal processing and electrical engineering are also represented with shares of 11.2% and 10.4% respectively. Furthermore, there is a small but still significant participation from the chemical sector (5.6%) and the glass, ceramics and plastics sector (4.0%). The category "Other" contains 16.8% of responses and includes participants from sectors such as the food industry, the construction industry and the service sector.

3.1 Utilized Systems for Production Control

In order to achieve a clearer overview of the current system landscape, we analyzed which production control systems are in place (survey question: Which systems are available for production control?).

The most frequently mentioned system for production control is Enterprise Resource Planning (ERP) with 88.8%. This reflects the central role of ERP systems in modern companies, as they help to integrate various business processes and optimize resource utilization. Spreadsheet analysis (Excel) follow in second place with 68.8%. This indicates that despite the availability of specialized systems, many companies still rely on traditional tools such as Excel to manage certain aspects of their production control. It also shows that simple and generally accessible tools often have a permanent place in the corporate landscape.

Operation Data Acquisition (ODA) is used by 55.2% of companies, highlighting its importance for real-time collection and analysis of store floor data. At 45.6% each, the Manufacturing Execution System (MES) and Machine Data Acquisition (MDA) are tied, reflecting the need for companies to monitor and control the production process in real time, collecting data directly from production machines.

Furthermore, 1.6% of the companies stated that they had no production management system at all. These are companies with very specialized production processes. Finally, 2.4% of respondents indicated that they use other systems not included in the list. This includes specialized and customized solutions developed for the specific needs of certain companies. The variety of systems mentioned illustrates that production control in many companies relies on a wide range of tools and solutions to meet their needs (Fig. 2). It is interesting to see that despite the use of other specialized systems, Excel still plays a significant role and is the dominant tool alongside ERP systems.

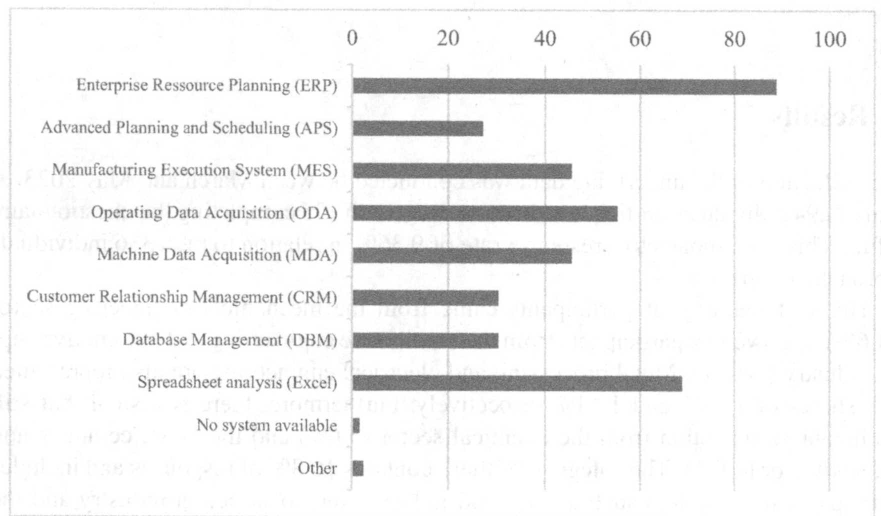

Fig. 2. Utilized systems for production control

3.2 Evaluation of the Systems Currently in Use

An analysis of the degree of fulfillment of certain functions was carried out to evaluate the performance and reliability of the systems. This analysis focused on how well the systems currently used in the company fulfill selected requirements (survey question: Evaluate the following statements with regard to the systems you currently use).

The systems provide reliable data: The majority of respondents (60.3%) indicate that the systems tend to provide reliable data (summed "strongly agree" and "agree"). However, a significant proportion of 34.7% are in the middle of the scale and perceive the reliability of the data as "partly agree". Only a small number of 4.96% are less satisfied with the reliability of the data.

The systems provide real-time data: With 52.4%, about half of respondents feel that systems are more capable of providing real-time data. A quarter (25.0%) are "partly agree" that the systems are fulfilling this demand in this regard. Slightly more than one in five (22.6%) find that the systems are less able to provide real-time data.

The systems generate accurate data: Half of the participants (50.4%) are satisfied with the accuracy of the data generated. One-third (33.3%) are in the middle of the scale and find that the data is "partly accurate". A small proportion of 16.3% state that the generated data is less accurate.

The systems have a common database: 43.8% of respondents feel that systems tend to share a common database. A proportion of 28.1% are in the middle of the scale with "partly agree" and see the common data base as partly available. The same proportion (28.1%) feels that the systems have less of a common database.

The systems are fast: A majority of participants (36.6%) feels that the systems are rather fast. Over a third (37.4%) are in the middle of the scale and see the speed of the systems as partly fast. A relevant proportion (26.0%) find that the systems are less fast.

The systems help to save time: With 55.6% over the half of respondents answered that the systems tend to help save time. Almost a third (31.5%) "partly agree" that the systems are helping in terms of saving time. A small proportion (12.9%) believe that the systems contribute less to saving time.

The systems are fail-safe: A significant number (51.6%) of participants feel that the systems are rather fail-safe. A substantial number of 29.5% "partly agree" that the systems' resilience. Almost one in five (18.9%) believe that the systems are less fail-safe.

The systems are available: Four out of five participants (80.5%) answered that the systems are somewhat available. A portion of 17.07% view the availability of the systems as "partly available". A very small proportion (2.44%) believe the systems are less available.

The systems are simple: With 41.46%, the majority perceive the simplicity of the systems as partly existing ("partly agree"). 23.58% of respondents consider the systems to be simple, but only 1.63% "strongly agree" at this point. A relevant proportion (35.0%) sees the systems as less simple.

The systems are understandable: The majority of the participants (46.3%) perceive the systems as "partly understandable". A significant portion (30.1%) find the systems to be somewhat understandable. A relevant proportion (23.6%) sees the systems as less understandable.

The survey results (Fig. 3) showing that the majority of participants are satisfied with various aspects of the systems, including data availability, real-time data, accuracy, availability, and time savings. However, some aspects such as fail-safety, simplicity and understandability offer room for improvement.

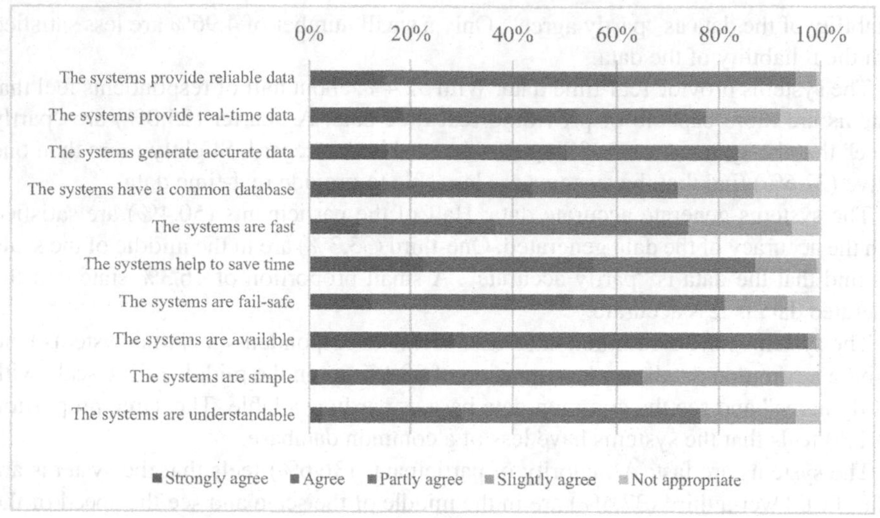

Fig. 3. Evaluation of the systems currently in use

3.3 Support of the Daily Activities by the Systems

To investigate how the systems support the daily work of users in terms of recognition and prioritizing problems and malfunctions as well as defining and prioritizing measures and solutions, the different categories were examined (survey question: To what extent are you supported by the systems in your daily work in the following points?).

Recognition of problems/malfunctions: The systems are predominantly rated positively when it comes to recognizing problems and disruptions. More than half of the respondents (50.8%) state that they are supported in this respect ("Strongly agree" and "Agree"). However, there is a significant 27.0% who find that the systems help them little or not at all in this area.

Prioritizing problems/malfunctions: Feedback is more mixed when it comes to prioritizing issues and problems. While 32.2% state that they are supported in this task ("Strongly agree" and "Agree"), 40.5% of respondents find that the systems provide them with little or no support in this area. Over a quarter (27.3%) are neutral on this.

Defining measures/solutions: Here it is clear that most respondents do not feel that the systems support them sufficiently. A majority of 60.7% say the systems help them little or not at all, while only 15.6% think they are supported in this regard. Almost a quarter (23.8%) have a neutral attitude.

Prioritizing measures/solutions: In this area, the picture is similar to that of defining measures. The majority (56.6%) find that the systems help them little to not at all, while only 21.31% state that they are supported, with this being fully true for none of the participants. About a quarter (22.13%) are neutral on this.

In summary, the survey results (Fig. 4) show that participants are supported to varying degrees in their daily work by the systems they use. The systems seem to be rated rather positively in supporting the identification of problems and faults as well as in prioritizing problems and faults. On the other hand, feedback shows that there is room for improvement in defining and prioritizing measures and solutions. To further optimize the work process and user effectiveness, systems in this area need to be specifically adapted and improved.

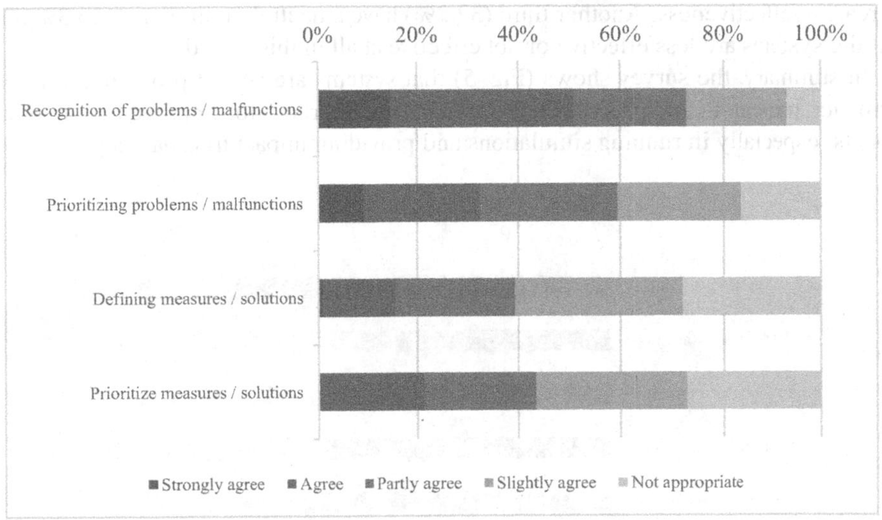

Fig. 4. Support of the daily activities by the systems

3.4 Performance of the Systems with Regards to Daily Work

For the evaluation of the system performance, an analysis of the degree of fulfillment with regard to daily support activities in form of selected services was carried out. (survey question: What services does your system provide in the daily work?).

Providing key figures: Most respondents' systems (69.1%) effectively provide key performance indicators, with 30.9% strongly agreeing and 38.2% agreeing. A smaller proportion (12.2%) find this to be less or not at all the case.

Performing simulations: Systems are not particularly strong here. The majority of respondents (61.0%) say the systems are less effective or not effective at all in terms of running simulations. Only 13.8% ("Strongly agree" or "Agree") see this as a strength of their systems.

Parameterization of characteristics: Opinions are mixed. While 21.9% "Strongly agree" or "Agree" that their systems do this effectively, 45.4% ("Slightly agree" or "Not appropriate") find the opposite. One-third (32.8%) have a neutral stance.

Transparency of the effects: The majority (51.6%) find that their systems are less effective or not effective in this regard. Only 25.4% ("Strongly agree" or "Agree") view their systems positively in this area.

Control of the process: For 44.7% ("Strongly agree" or "Agree") of participants, their systems help them control the process. A proportion of 30.9% ("Slightly agree" or "Not appropriate") do not see it this way, and almost a quarter (24.4%) have a neutral attitude.

Increase of reactivity: While 34.5% ("Strongly agree" or "Agree") think their systems help increase reactivity, there is a larger proportion (37.8%) with a neutral opinion. 27.7% think the systems are less effective or not effective at all here.

Increase of effectiveness: A majority (41.5%) believe their systems are effective at increasing effectiveness. Another third (32.2%) have a neutral opinion, and 26.3% find that the systems are less effective or not effective at all in this regard.

In summary, the survey shows (Fig. 5) that systems are best at providing key performance indicators and process control. There is clear room for improvement in other aspects, especially in running simulations and providing impact transparency.

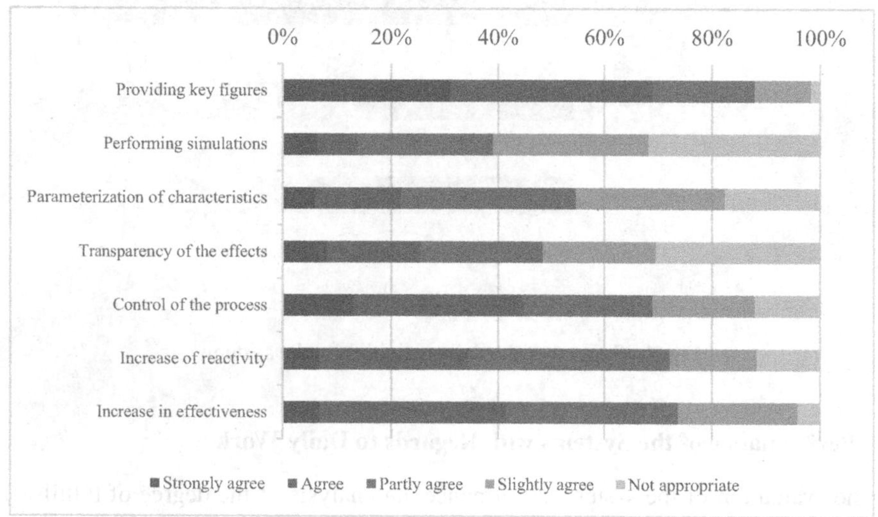

Fig. 5. Performance of the systems with regards to daily work

3.5 Provision of System Functionalities

The provision of system functionalities was determined with regard to requirements (target situation) and the degree of fulfillment (actual situation) in order to find out how the system supports the challenges that arise (survey question: Which functionalities does the system provide or should the system provide?).

The representation of data/information in real-time exhibits a difference of 45% between the demand (target situation), and the actual degree of fulfillment (actual situation). This is noteworthy, especially considering that this function has the highest demand at 93%. Similar discrepancies are found in functionalities for processing of data or information and the digital image of data or information in real-time, which have differences of 40% and 48%, and demands at high levels of 90% and 88%, respectively. In contrast, the 3D mapping of the production environment shows the smallest difference of 18% but also has the lowest demand at 34%.

A significant discrepancy between the target and actual situation is also noticeable in other functionalities. Proactive decision support and the integration of intelligent and self-learning algorithms have the largest differences, at 57% and 56% respectively. The demand for proactive decision support is at 81%, and for the integration of intelligent algorithms, it is at 70%. Additionally, the functionalities for the integration of realistic simulations and the autonomous control of subtasks should be mentioned, with differences of 49% and 52%, and demands at 71% and 74%, respectively.

In summary, the results (Fig. 6) indicate that the listed system functionalities are of great importance to the vast majority of respondents. However, the actual status shows that the majority of users are not satisfied with the current implementation of these functionalities. This underscores a significant need for improvement. If the systems were capable of better meeting these needs, it could lead to a substantial increase in user satisfaction.

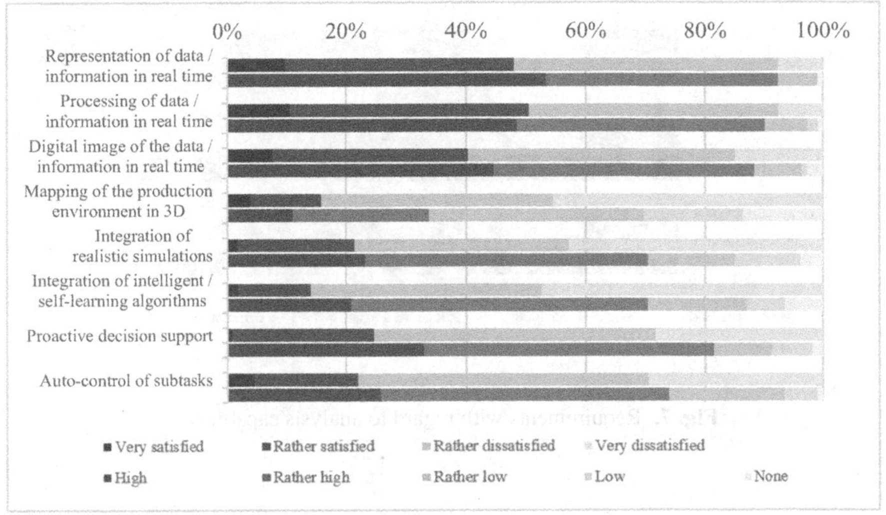

Fig. 6. Provision of system functionalities

3.6 Requirements with Regard to Analysis Capability

In order to find out which insights the system provides for decision preparation, the support of the system in terms of analysis capability with regard to the need (target situation) and the degree of fulfillment (actual situation) was determined (survey question: How does the system support or should support the analysis capability?).

In terms of analytical capability, the highest demand (91%) is for descriptive analysis (what happened?). The difference between the need and the degree of fulfillment is 46%. The diagnostic analysis (why did it happen?) shows a need of 86% and includes the largest proportion of 47% who consider the functionality to be very important. The difference between the target and actual situation is 68%.

The greatest discrepancy between the need and the actual degree of fulfillment is in the predictive analysis (what will happen?) with a difference of 74%. This form of analysis has the second-highest demand at 90%. In the prescriptive analysis (what needs to be done?), there is a difference of 70% between the targeted and achieved performance levels. The need for prescriptive analysis is indicated by users at 86%.

Overall, there is a discrepancy in all areas (Fig. 7) between the need for these functions and satisfaction with the current implementation in the system. The descriptive analysis is the only one which shows satisfaction regarding the degree of fulfilment with 46%, of which only 6% are very satisfied.

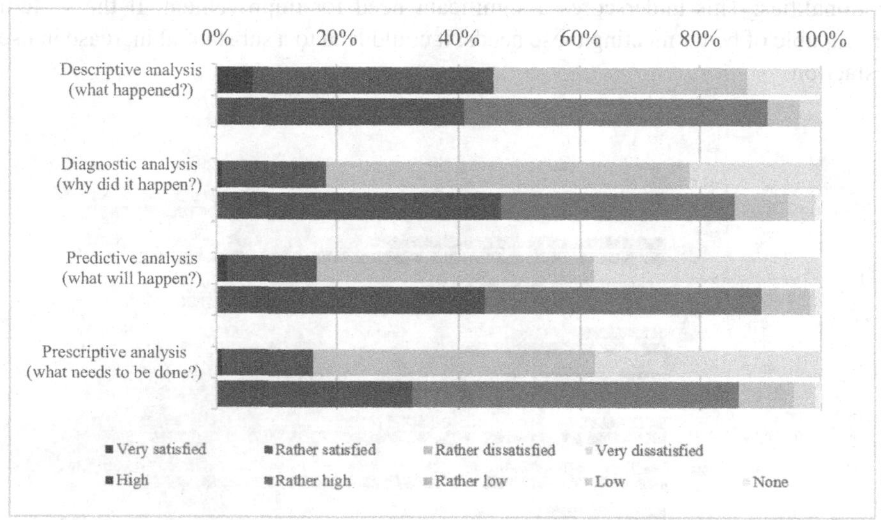

Fig. 7. Requirements with regard to analysis capability

3.7 Requirements with Regard to Decision Support

The requirements (target situation) and the degree of fulfillment (actual situation) with regard to decision support were determined in order to find out how the system supports the decision-making process for target/performance adjustments (survey question: What does the system provide or should provide in terms of decision support?).

The greatest need for decision support is in the analysis of data and results, at 96%. Closely following are the functionalities control of circumstances and recommendation of actions, at 89% and 87% respectively. Particularly noteworthy is that the functionality of recommending actions exhibits the largest discrepancy between the target and actual situation, at 66%. The functionalities of analyzing data / results and control of circumstances have a similar pattern, with the discrepancies between need and fulfillment both at 47%.

Furthermore, it is evident that the functionality of deciding on actions has a need of 61%, while the lowest demand, at 58%, is for initiating an action. The functionality of controlling processes is stated to have a need of 86%. For the functionalities deciding on actions and controlling processes, the discrepancy between the target and actual situation is 40% and 47% respectively. The smallest difference is recorded at 37% for the functionality of initiating an action.

The results (Fig. 8) clearly indicate that there is a significant need for decision support in various functional areas, especially in the analysis of data and results, the control of circumstances, and the recommendation of actions. However, it is noticeable that in many of these areas, there are significant discrepancies between the demand (target situation) and the current level of fulfillment (actual situation), with the function of recommending actions showing the largest difference. These discrepancies underscore the need to improve and further develop system functionalities to better meet user needs and enhance the efficiency of decision-making processes. In particular, areas with the highest demand and the largest differences should be treated as a priority.

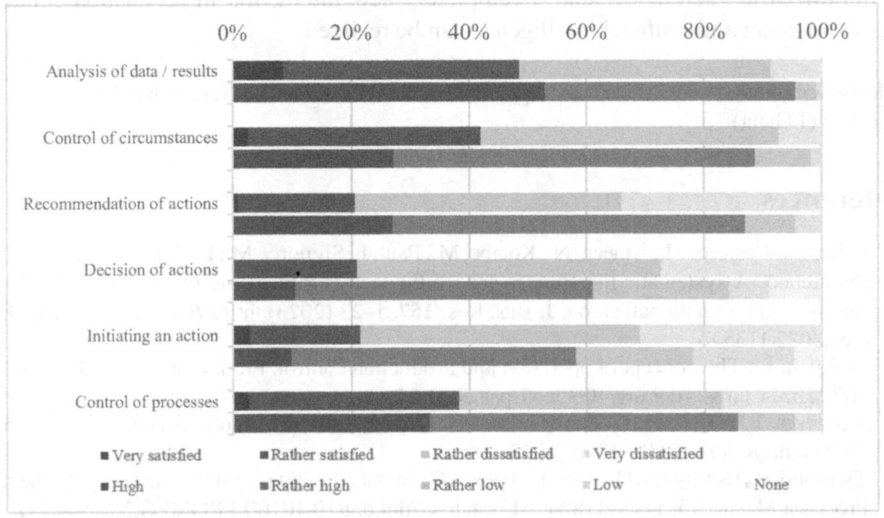

Fig. 8. Requirements with regard to decision support

4 Conclusion

The systems currently in use assist in identifying problems or disturbances and primarily provide decision-makers with basic functions such as supportive key figures. Regarding the performance of the systems, there is particularly potential in areas related to the definition and prioritization of measures and solutions. Users and operators compensate for missing or inadequate features of the systems, leading to a significant portion of the decision-making basis being developed, structured, and interpreted by humans.

The area of representation and processing, as well as the digital image of data and information in real-time, records the highest demand for necessary system functionalities. The most pronounced discrepancy between need and current implementation status is evident concerning proactive decision support and the integration of intelligent, self-learning algorithms.

In terms of analytical capability, there is potential in particular in areas that extend descriptive analysis (what happened). The degree of fulfillment does not meet the requirements for both diagnostic analysis (why did it happen) and predictive analysis (what will happen).

The expectations of the respondents regarding the decision support to be provided by the system are diverse. Of particular importance is the expansion of data and results analysis to enable the surveillance of circumstances and the recommendation of actions.

The survey results present a multifaceted picture of the current sentiment regarding the status quo of industrial systems for production control. While the identified potentials and needs are significant, they can be used as an outline for the transformation into a digital environment which is supporting the operational challenges. Especially through targeted efforts by the integration of artificial intelligence, the identified challenges can be optimized in such a way that a completely new interaction in production control between human and artificial intelligence can be realized.

Acknowledgments. This paper was supported by the BTU Graduate Research School (Conference Travel Grant).

References

1. Feliciano-Cestero, M., Ameen, N., Kotabe, M., Paul, J., Signoret, M.: Is digital transformation threatened? A systematic literature review of the factors influencing firms' digital transformation and internationalization. J. Bus. Res. **157**, 1–22 (2023). https://doi.org/10.1016/j.jbusres.2022.113546
2. Wolniak, R.: The concept of operation and production control. Prod. Eng. Arch. **27**(2), 100–107 (2021). https://doi.org/10.30657/pea.2021.27.12
3. Sturgeon, T.J.: Upgrading strategies for the digital economy. Glob. Strateg. J. **11**, 34–57 (2021). https://doi.org/10.1002/gsj.1364
4. Tjahjono, B., Esplugues, C., Ares, E., Pelaez, G.: What does industry 4.0 mean to supply chain? Procedia Manuf. **13**, 1175–1182 (2017). https://doi.org/10.1016/J.PROMFG.2017.09.191
5. Wagner, T., Herrmann, C., Thiede, S.: Industry 4.0 impacts on lean production systems. Procedia CIRP **63**, 125–131 (2017). https://doi.org/10.1016/j.procir.2017.02.041
6. Hofmann, E., Rüsch, M.: Industry 4.0 and the current status as well as future prospects on logistics. Comput. Ind. **89**, 23–34 (2017). https://doi.org/10.1016/j.compind.2017.04.002
7. Li, L., Rong, S., Wang, R., Yu, S.: Recent advances in artificial intelligence and machine learning for nonlinear relationship analysis and process control in drinking water treatment: a review. Chem. Eng. J. **405**, 1–17 (2021). https://doi.org/10.1016/j.cej.2020.126673
8. Andronie, M., Lăzăroiu, G., Iatagan, M., Uță, C., Ștefănescu, R., Cocosatu, M.: Artificial intelligence-based decision-making algorithms, internet of things sensing networks, and deep learning-assisted smart process management in cyber-physical production systems. Electronics **10**(20), 1–24 (2021). https://doi.org/10.3390/electronics10202497
9. Schmid, S., Winkler, H.: Die Produktionssteuerung im digitalen Zeitalter aus Sicht der industriellen Praxis. IKMZ, Cottbus (2023). https://doi.org/10.26127/BTUOpen-6500
10. Dieckmann, A.: Empirische Sozialforschung. Rowohlt Taschenbuch Verlag, Reinbek bei Hamburg (2013)
11. Yin, R.K.: Case Study Research Design and Methods. Sage Publications, Thousand Oaks (2014)
12. Kromrey, H.: Empirische Sozialforschung - Modelle und Methoden der standardisierten Datenerhebung und Datenauswertung. Lucius & Lucius, Stuttgart (2009)

Process Improvement of Taping for an Assembly Electrical Wiring Harness

Ikhlef Jebbor[1(✉)], Youssef Raouf[1], Zoubida Benmamoun[2], and Hanaa Hachimi[1]

[1] Laboratory of Advanced Systems Engineering, National School of Applied Sciences, Ibn Tofail University, Kenitra, Morocco
jebbor.ikhlef@uit.ac.ma

[2] Faculty of Engineering, Liwa College of Technology, 31009 Abu Dhabi, United Arab Emirates

Abstract. The production of automotive wire harnesses now requires a significant amount of manual labor. Even so, a greater level of automation is required due to present and future application demands such as the miniaturization of electronic components, the monitoring of process parameters, the growing need for paperwork for processes, and the rise in payments. Technology helps manufacturing organizations in the most important aspects of the design-to-manufacturing process. This is relevant to the wire harness sector, which is a crucial part of industrial automotive manufacturing. The automotive wiring harness suppliers designed Computer-Aided Design technologies (CAD) to support wire harness assembly operations and design work. Even with the application of these techniques, engineers will still have to do trial-and-error work to find effective assembly techniques. This study presents a new approach to optimize the wire harness assembly procedures without focusing on trial and error or the experience of experienced engineers to develop operational assembly processes. The most crucial and challenging step in the assembly process sequence is taping routed cables. The taping process's complexity is mostly determined by how the jig is set up on the workstation and the tape method. As a result, the suggested technique models and optimizes the tape direction and jig arrangement.

Keywords: Electrical wiring harness · Industrial engineering · Lean Manufacturing · Layout jigs · Optimization of taping

1 Introduction

1.1 Wiring Harness Evolution

The wire harness business has made significant advancements in recent years due to the automobile industry's ongoing growth. For most of this century, the diversity of automobiles has been continuously rising [1]. Wiring harness size and total weight have increased together with the increasing number of electrical components [2]. If we take into account all of the advancements made in the automobile industry in 2008, software accounted for 80% of the total and electronic components for 90% [3]. The parallels

S.-H. Sheu (Ed.): ICIEA-EU 2024, LNBIP 507, pp. 35–48, 2024.
https://doi.org/10.1007/978-3-031-58113-7_4

between technology and automotive technology are growing [4]. A vehicle built in the 1960s was mostly made of mechanical parts. With ten million lines of code and 40 electrical control units, electrical wire harnesses have a combined length of almost 8000 m of cable [5].

Comparably, hybrid vehicles need additional safety features to prevent damage from short circuits because they contain further cables [6]. Because these security features duplicate electrical networks and electronic control units, there are more wires involved. An electric or hybrid vehicle's connections and wiring harness components must be capable of operating at their best without running the danger of malfunctioning or failing. This is the rationale for the installation of many secure systems in every wire harness.

Cars that run on electricity or gasoline have a chance to be sustainably good [7, 8]. Since electrical components in cars have increased in significance over the years, it is difficult to reduce the weight of wire harnesses [9, 10]. Over twenty kg of wires are present in medium-sized cars, and cutting the mass of the wiring harness minimizes fuel consumption [11]. Designers therefore have to make adjustments and figure out a solution to make the machine as light as feasible under a more complicated electrical wiring scenario. Therefore, wire harnesses must provide techniques that help reduce the effects of more technology on weight, cost, complexity, and packing area. Aluminum wire is being used in place of the copper wires that were used earlier. The cars are lighter because to the aluminum wire harness [12]. Component miniaturization is another technique used to enhance wiring harness density since it reduces costs, increases efficiency, plus reduces weight [13].

An electronic wiring diagram is made, showing all of the connections between the various electronic parts, along with the voltages, currents, and consequent wire widths that are required. Twisted wire pairs, as opposed to single ones, must be used in critical controlled connections to minimize electromagnetic interference on electrical signals [14].

A 3D CAD plan is generated as well, which considers several impacting aspects such as relative motions of the components, assembly areas cutting edge various degrees of temperature, material resistance, and installation demands, as well as the component's positions. These contributing elements typically result in the need for special-purpose components and an altered layout. A thorough 2D engineering drawing is produced using the wiring schematic and the 3D CAD layout as references. The regular installation board that is used to build the wire harness has been assigned using this 2D design.

1.2 Assembly and Tape Harness

Wire harnesses are found in every plane, jet, and rocket in the sky as well as beneath the hood of every vehicle and truck on the road. Cable assemblies or harnesses for wires are other names for them. An organized collection of cables or wires wrapped to create an integrated unit is described by all of these words. This integrated device connects the many components of the automobile, truck, or airplane by transmitting signals and electrical power. For almost every contemporary car to operate properly, wire harnesses are essential.

Modern cars are fitted with electronic parts that do several additional tasks besides the standard ones of moving, turning, and stopping. In order to manage these electronic parts and send power and messages to every area of the car, wiring harnesses are essential. They are vital parts, similar to the blood arteries and nerves in the human body, and consist of connections, terminals, clamps, sheaths, and other components around a core of electrical wires.

For many years, research has been conducted to assist the design of manufacturing processes, since companies have long been concerned with the efficiency of their operations [15–17]. Due to the few studies addressing this topic, we concentrate on the assembly procedures of automobile wire harnesses in this research.

Cables or wires that carry electrical power or communications are assembled into a wire harness. The cables are linked with tape, tube, sleeve, or other methods to preserve them and shorten the time needed to install a harness on a device. Many different sorts of items employ wire harnesses. One of the most notable examples of this is the automobile. In the automotive industry, using a workstation fitted with assembly jigs shown in Fig. 1, an electrical wire is produced by hand employing the following methods:

- After cutting of wire with the required length, their ends are terminated (see Fig. 2);
- Through jigs, the cables are routed. Jigs, which resemble forks, are used to keep routed cables in place during assembly procedures (see Fig. 3);
- Connector housings are put into the cables' ends (see Fig. 4);
- In the end, tape is used for joining cables (see Fig. 5).

Even though today's automotive production methods are greatly automated; hand manufacturing is still required for wiring harnesses. The complexity of the several processes involved and the challenge of automating them are the reasons for this.

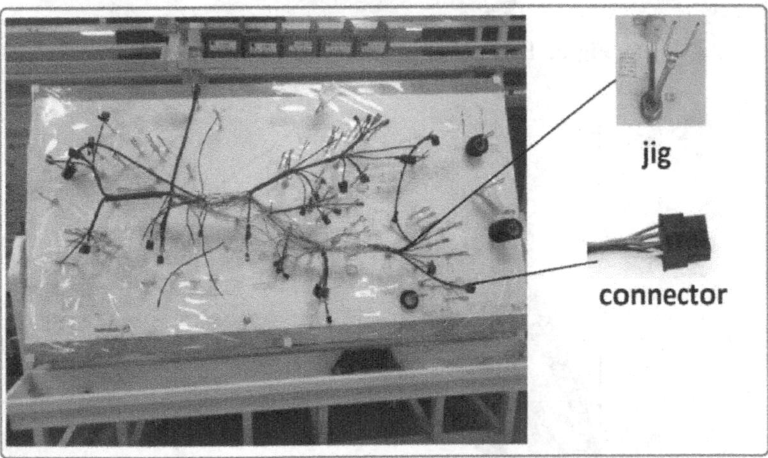

Fig. 1. A workstation for an assembly of automotive electrical wiring harness.

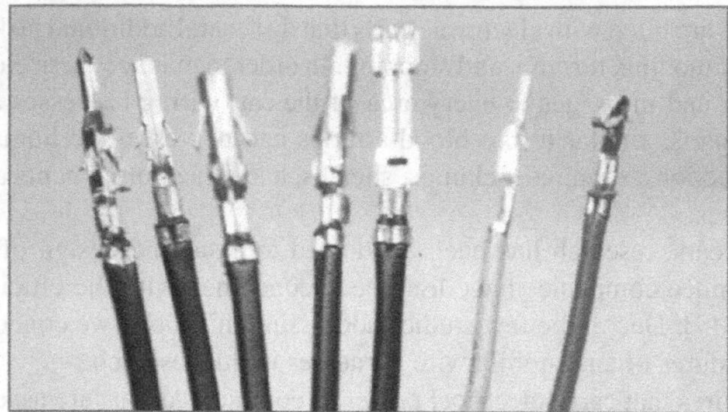

Fig. 2. Example of wires cut, and their ends terminated.

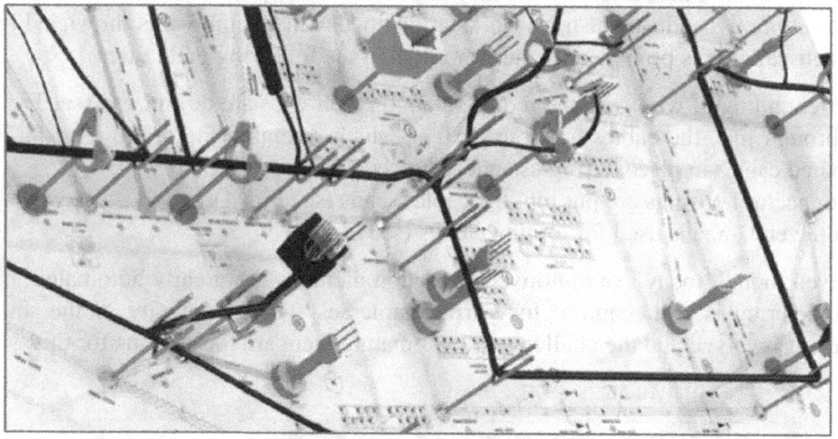

Fig. 3. Example of cables routed Through jigs

Fig. 4. Example of cable terminated inserted into Connectors.

Three different kinds of materials are often utilized to make the outside cover:

1) Corrugated pipe that exhibits good resistance to fire, abrasion, and water, as well as exceptional resilience to high temperatures. The range of temperature resistance is between −40 °C and 150 °C. The corrugated tubing used in engine wiring harnesses should be standard.
2) PVC tube (Poly Vinyl Chloride tube). PVC tubes are more flexible and have a higher resistance to bending strain than corrugated tubes. The temperature resistance is, however, somewhat lower (around 80 °C). It can be applied to the branches of the wire harnesses' front section.
3) Adhesive tape: used in somewhat safe working situations, such as on door and dashboard wires.

Current wire harness design functions are supported by Computer-Aided Design (CAD) And Computer-Aided Manufacturing (CAM) [18, 19]. Additionally, specialized software to aid in wire harness design [19, 20] has been created. Numerous techniques and technologies, including CAD/CAM-based systems [21–23], virtual reality-based systems [24, 25], process planning [26, 27, 28, 29, 30], and expert systems [31, 32], have been developed to help the design of wire harness manufacturing procedures. Designers must still develop assembly procedures by trial and error, even with applying these techniques. Therefore, the effectiveness of assembly processes is highly dependent on the engineering level, and investigating good processes for production necessitates extensive procedures. Therefore, to develop efficient processes of electrical wiring without experiencing trial with error, an optimization-based design strategy is required [33, 34]. The study focuses on the step when an operator of tape-routed cables is the most crucial and difficult task in the building sequence. The complexity of the tape procedure changes but the workspace jig's arrangement and the tape path have a major influence. As an example, impact from branches makes it challenging to tape each branch when there is too little angle or when two branches of the wire harness intersect. Thus, a genetic algorithm is used to simulate and optimize both a jig arrangement and a tape path in the suggested strategy [35–37]. Regarding the GA fitness operation, as working time is an especially crucial factor, the amount of time spent working is determined by using a taping route plus a jig's arrangements. This is handled as the fitness function [38, 39].

This paper is structured according to the format for the remainder of it. The suggested method's specifics are explained in Sect. 2. In Part Two, a wire harness on a workstation, a set of jigs, and a path for tape tasks are evaluated and modeled. Then, individuals, fitness functions, and constraint conditions are established in order to optimise these through the use of a genetic algorithm [40, 41].

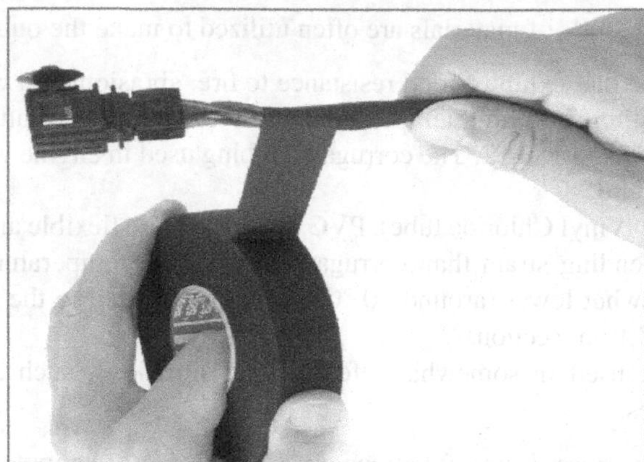

Fig. 5. The tape operation for an assembly of automotive electrical wiring harness.

2 Method

2.1 Overview of the Suggested Technique

This article concentrates on the step when an employee tapes the routing wires, even if a wire harness is created using the procedures specified in Part 1. For this reason, the procedure is the most charging on a worker's body, and its jobs significantly impact the time spent in assembly. A jig's arrangement on a workstation and a tape path has a significant impact on its difficulties. Regarding the jig design, branch interference makes it challenging to tape every branch if there is too little space between two wiring branches or if two branches intersect. Therefore, it is best if there is a big enough angle (at least 90 degrees or more) between each of the branches, and branch intersections should be minimized. But it's hard to put every branch on a workbench because of its small dimensions and the fact that modern cars often have wire harnesses with plenty of branches. It is necessary to investigate the arrangement that requires the least amount of working time for taping procedures. The procedure of tape involves two tasks: the practical tape of the branch and the movement of the operator's hands from one area to another without performing any labor. A traversable tape path is unlikely. A tape route does not influence the original task's distance, whereas it has a significant impact on the second task's distance. The route that reduces its distance must be investigated as the last obligation is pointless labor. Furthermore, because the distance between branches depends on each branch's placement, which is established by a jig's layout, the final task's distance depends not only on a tape path but also on a jig arrangement on a workstation. There is additionally a chance that the movement distance without any work may be reduced even more if the jig arrangement is optimized by assessing both the overall taping time and the movement distance without any task. For that reason, the article concurrently optimizes the path of tape tasks and the design of the jigs.

In the study, the best possible design issue is summarized as follows: Two factors that are considered as valuation aspects are the amount of time required to tape every branch and the distance left unemployed, while the workstation jigs configuration and a taping path task are treated as factors influencing the design. They serve as the foundation for the problem's fitness purpose. We offer a genetic algorithm to address the issue ideal layout challenge.

Part 2 is structured as follows for the remainder of it. The value aspect of the design was identified initially, and the workstation configuration comprising jigs and a wiring harness was modeled. Next, the value feature of the path of tape jobs is established and modeled. Persons, constraining conditions, fitness functions, and persons are at last specified as a way to optimize them utilizing genetic algorithms.

2.2 Workstation Set up a Wire Harness and Jigs Model

The jig system and the wiring harness are displayed on a workstation in Fig. 6. A workstation is represented by an outer rectangle. The harness's branching point or endpoint is Node N. There are jigs at every node. A circle in the illustration represents a node. A connector, indicated by a rectangular shape, is fixed to the end of the cable. Edge E is the portion of the harness that connects both nodes. The edge E_j has the following measurements: l_j, d_j, and θ_j for length, diameter, and angle, respectively. The angle θ_j represents the angle formed in an opposing direction by a line that is horizontal and the edge Ej. The term "The main border" refers to the edge that has the greatest number of wires, while "The branch side" refers to all other edges. A reduced line represents a branch edge, while a large line indicates the principal edge. The wiring harness arrangement, as well as the diameter and length of every edge, will be defined during the optimization step. On the other hand, every edge's angle serves as a design variable.

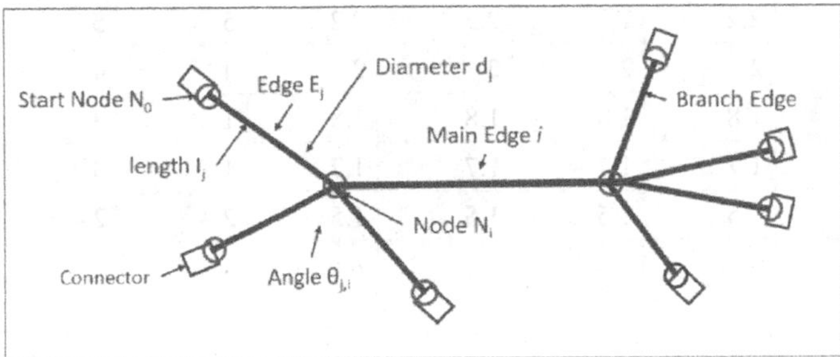

Fig. 6. Model of a workstation assembly electrical wiring harness.

2.3 Total Tape Time Determination

The amount of time required to tape each side, or total taping time T, is used to value the jigs arranged on a workspace. The following formula defines the sum of the tape duration time T:

$$T = \sum\nolimits_{i=1}^{N} t_i \omega_i \tag{1}$$

Although $\mathbf{t_i}$ is the average working time required for taping every edge and $\mathbf{E_i}$: the shortest timing required. If there are no surrounding boundaries or other obstacles, then the Table 1 determines $\mathbf{E_i}$ using diameter **di** and length **li**.. $\mathbf{t_i}$. The tape edge Ei's difficulty rating is represented by $\boldsymbol{\omega_i}$. As previously said, taping edges that have a tiny angle within them gets challenging. Table 2 computes $\boldsymbol{\omega_i}$ by taking the angle between diameter di and the neighborhood edge. The less large edge is chosen after calculating the angles with edge Ei and its surrounding edges ($\boldsymbol{\theta_{j,i}}$).

Table 1. Normal working time calculation table.

(a)	Diameter [mm]						
length [mm]	9	10	11	12	13	14	...
20–40	1.4	1.5	1.7	1.7	1.7	1.7	
41–60	2	2.1	2.2	2.2	2.2	2.2	
70–80	2.4	2.5	2.7	2.7	2.7	2.7	
81–100	3	3.2	3.2	3.3	3.3	3.3	
101–120	3.4	3.5	3.8	3.8	3.8	3.8	
...							

Table 2. Difficulty calculation table.

(b)	Diameter [mm]						
Angle [deg]	9	10	11	12	13	14	...
15	2.2	2.2	2.2	2.2	5	5	
30	2	2	2	2	4	4	
45	1.8	1.8	1.8	1.8	1.8	3	
60	1.7	1.7	1.7	1.7	1.7	3	
75	1.5	1.5	1.5	1.5	2.4	2.4	
...							

2.4 Scheduling Tasks in Sequence According to a Travelling Salesman Problem

The following details, when combined with the order and direction of taping on each edge, may determine the full tape path. The traveling salesman issue is used to describe the sequence of edges that need to be taped. Figure 4 illustrates this concept by using edges to represent cities and the sequence in which the edges should be taped to represent a salesman's trips to the locations. The pattern P indicates the order of numbers that indicates which edges need to be taped. P is {1, 4, 5, 7, 8, 3, 6, 2} in the instance of Fig. 7.

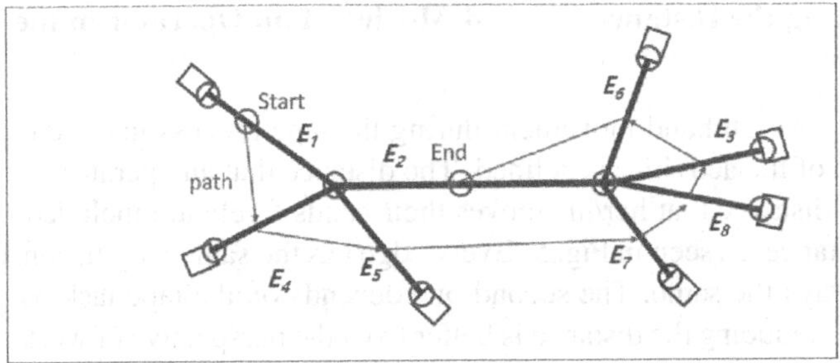

Fig. 7. A traveling salesman issue is used to indicate the sequence of edges that need to be taped.

On the other hand, binary coding indicates which way each edge should be taped. At both ends of every edge are a pair of nodes. "Start or Begin point" refers to the node that is closest to node N0, while "Endpoint" refers to another node. When the direction is zero, An operator tapes the area with tape between the beginning and the end. Assuming that direction of 1, an operator tapes the area between the finish and the beginning. The set of binary codes that correspond to the directions on each edge is called D. The direction that has to be taped is shown in Fig. 8.

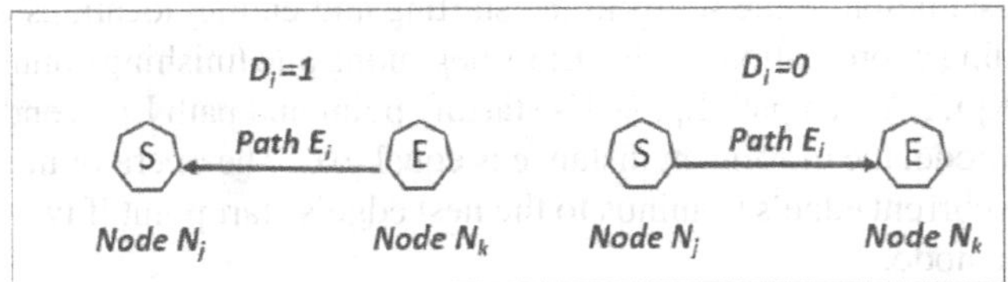

Fig. 8. Taping direction.

Each step of the recording procedure is determined where two parameters of data, D and P, are mentioned. Figure 9 shows the method of determining the path.

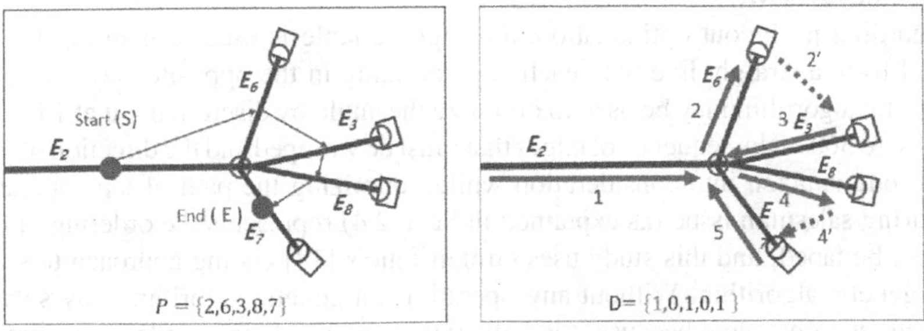

Fig. 9. Flow of identifying the route of the taping process.

2.5 Calculating the Distance of Hand-Moving of an Operator in the Absence of Activity

The operator's overall hand movement during the tape process may be measured after the entire path of the activities is defined. The distance that an operator physically tapes edges and the distance that her/his moves their hands freely are included in the overall movement distance, as seen in Fig. 5. Every edge has the same length, consequently, the original is always the same. The second one depends on the tape task path and the jig configuration. Reducing the distance is better from the perspective of working efficiency while operator moving their hands when not engaged in any activity is just waste. As such, the amount of space a worker's hands move when not performing a job is denoted by the letter M and used to assess a taping path. The following formula defines M.

$$M = \sum_{i=1}^{N-1} \left\| R_{P_i,k} - R_{P_{i+1},m} \right\| \tag{2}$$

$$\begin{cases} D_{P_I=0 \to k=2} \\ D_{P_I=1 \to k=1} \\ D_{P_{I+1}=0 \to m=1} \\ D_{P_{I+1}=1 \to m=2} \end{cases} \tag{3}$$

R is a representation of the set from the starting and ending locations of each edge. The corresponding coordination of edge Epi's beginning and finishing points is displayed by Rpi,1 with Rpi,2. When path Epi + 1's starting point and path Epi's ending point are identical to the node, the movement distance is equal to 0. The operator must move their hands from the current edge's terminus to the nest edge's start point if two locations are not on the same node.

2.6 Optimization of Moving Distance and Total Tape Time T Without Any Task M

This research uses a genetic algorithm to determine the jig arrangement and the tape assignments path with the lowest possible total tape time T and the movement range without tape tasks M.

Regarding jig layout optimization, a design variable is used to manage the angle obtained from a straight line with each edge pointing in the opposite way orientation. The genetic algorithm may be used to optimize the angle by discretizing it at 15 degrees from zero to 360°. The sequence of edges that must be wrapped and the direction of taping on everyone is taken into consideration while optimizing the path of tape operations. The touring salesman issue (as explained in Sect. 2.4) represents the ordering of edges that must be taped, and this study uses Grefenstette's [18] coding approach to solve it with a genetic algorithm. Without any operations, a genetic algorithm may solve the taping direction problem because it is only expressed as a binary code. To summarize, any person has 3 chromosomal: the angle is formed by all edges pointing in different directions and a straight line. Manner, the edges that need to be taped in the correct sequence, and the direction of taping on each edge. The total working time (T_{total}), which is determined by the equation below, serves as the genetic algorithm's fitness function.

$$T_{total} = T + \frac{M}{400} \tag{4}$$

Here M is the traveling distance without any job and T is the entire amount of time spent taping, as explained in Sects. 2.3 and 2.5. When there is no task at hand, the operator moves their hands on the workstation at a pace of 400[mm/s].

Regarding constraint factors, a number of them are set up as follows because of electrical wiring design needs and assembly process effectiveness. An operator's fitness function incurs a penalty if it deviates from these restriction criteria.

- A linear arrangement of the major edges is preferred.
- Edges shouldn't ever overlap one another.
- Edges shouldn't extend outside the workbench's limits.
- Each node's edge adjacency relationship is predetermined and shouldn't be modified.

3 Conclusion

This study focuses on how an operator tapes the wires for an automotive wire harness, and it recommends integrating the optimization of taping operations and jig layouts on a workstation to achieve ideal production methods. The vehicle wiring harness fitted in a specific real automotive is used to demonstrate the efficacy of the suggested approach. The findings demonstrate that while the suggested strategy may optimize the tape task's route and jig arrangement concurrently, almost equivalent outcomes will be achieved by applying both sorts of optimizations separately and consecutively. As a result, depending on how the wiring structure is configured, the user should choose between combined and sequential optimization.

In terms of future development, we propose to make a case study for the suggested approach used on the real automotive wiring harness produced, and we want to focus on both the design of the electrical harness itself and the development of the method and the process and to investigate an individual, a process, a situation, or to conclude important concepts and findings that aid in forecasting future trends, reveal before unnoticed problems that have practical applications, and/or offer a way to more clearly comprehend a significant study subject. It is anticipated that assembly workflow effectiveness would increase further by taking into consideration the assembly procedures while designing a wire harness.

References

1. Fisher, M.L., Ittner, C.D.: The Impact of Product Variety on Automobile Assembly Operations: Analysis and Evidence. Wharton School, Philadelphia (1996)
2. Trommnau, J., Kühnle, J., Siegert, J., Inderka, R., Bauernhansl, T.: Overview of the state of the art in the production process of automotive wire harnesses, current research and future trends. Procedia CIRP **81**, 387–392 (2019)

3. Hardung, B., Kölzow, T., Krüger, A.: Reuse of software in distributed embedded automotive systems. In: Proceedings of the 4th ACM International Conference on Embedded Software, pp. 203–210, Pisa, Italy, 27–29 September 2004
4. Rivero, A.A.L.: From complex mechanical system to complex electronic system: the case of automobiles. Int. J. Automot. Technol. Manag. **14**(1), 65–81 (2014)
5. Frigant, V., Miollan, S.: The geographical restructuring of the European automobile industry in the 2000s (2014)
6. Dijk, M., Kemp, R.: A framework for understanding product market innovation paths–emergence of hybrid vehicles as an example. Int. J. Automot. Technol. Manag. **10**(1), 56–76 (2010)
7. Oliver, J.D., Rosen, D.E.: Applying the environmental propensity framework: a segmented approach to hybrid electric vehicle marketing strategies. J. Mark. Theory Pract. **18**, 377–393 (2010)
8. Tran, D.-D., et al.: Thorough state-of-the-art analysis of electric and hybrid vehicle powertrains: topologies and integrated energy management strategies. Renew. Sustain. Energy Rev. **119**, 109596 (2020)
9. Loos, F., Ließ, H.D., Dvorsky, K.: Simulation methods for heat transfer processes in mechanical and electrical connections. In: 2011 1st International Electric Drives Production Conference, pp. 214–220. IEEE, September 2011
10. Rius Rueda, A.: A Novel Optimization Methodology of Modular Wiring Harnesses in Modern Vehicles: Weight Reduction and Safe Operation. Universitat Politècnica de Catalunya, Barcelona, Spain (2017)
11. Koch, S., Antrekowitsch, H.: Aluminum alloys for wire harnesses in automotive engineering. BHM Berg-und Hüttenmännische Monatshefte **152**, 62–67 (2007)
12. Yamano, Y., et al.: Development of aluminum wiring harness. SEI Tech. Rev. **73**, 73–80 (2011)
13. Fernandes, M.M., de Almeida, I.A., Junior, H.M.: Automotive miniaturization trend: challenges for wiring harness manufacturing. No. 2010-36-0160. SAE Technical Paper, 2010
14. Reif, K.: Bosch Autoelektrik und Autoelektronik: Bordnetze, Sensoren und elektronische Systeme; mit 43 Tab. 6th edn. Wiesbaden: Vieweg+Teubner Verlag/Springer Fachmedien Wiesbaden GmbH Wiesbaden (2011). https://doi.org/10.1007/978-3-8348-9902-6
15. Benbrahim, H., Hachimi, H., Amine, A.: Deep transfer learning pipelines with apache spark and keras TensorFlow combined with logistic regression to detect COVID-19 in chest CT images. Walailak J. Sci. Technol. (WJST) **18**(11), Article 13109, 14 (2021)
16. Jebbor, I., Benmamoun, Z., Hachimi, H., Raouf, Y., Haqqi, M., Akikiz, M.: Improvement of an assembly line in the automotive industry: a case study in wiring harness assembly line. In: Tang, L.C. (ed.) Advances in Transdisciplinary Engineering, pp. 62–71. IOS Press, Amsterdam, The Netherlands (2023)
17. Jebbor, I., Benmamoun, Z., Hachimi, H.: Optimizing manufacturing cycles to improve production: application in the traditional shipyard industry. Processes **11**, 3136 (2023). https://doi.org/10.3390/pr11113136
18. CATIA. http://www.3ds.com/. Dassault Systèmes
19. O'B Holt, P., et al.: Immersive virtual reality in cable and pipe routing: design metaphors and cognitive ergonomics. ASME J. Comput. Inf. Sci. Eng. **4**(3), 161–170 (2004). https://doi.org/10.1115/1.1759696
20. Sekine, T., Ito, T., Usuki, S., Miura, K.T.: Electric property analysis and wire placement optimization of automotive wire harness. In: 2021 IEEE International Joint EMC/SI/PI and EMC Europe Symposium, Raleigh, NC, USA, p. 189 (2021). https://doi.org/10.1109/EMC/SI/PI/EMCEurope52599.2021.9559207

21. Ruan, J., Zhou, X.: Research on CAD/CAPP integrated system for automobile wiring harness. In: 2011 Second International Conference on Mechanic Automation and Control Engineering, Inner Mongolia, China, pp. 5146–5149 (2011). https://doi.org/10.1109/MACE.2011.5988240
22. Kim, J.H., Lee, J.C., Park, J., Lim, H.: Electronic control unit modeling for automotive wire harness simulation using the Capital Harness system. Proc. Inst. Mech. Eng. Part D: J. Automob. Eng. **225**(3), 294–304 (2011). https://doi.org/10.1177/2041299110393215
23. Ritchie, J.M., Robinson, G., Day, P.N., et al.: Cable harness design, assembly, and installation planning using immersive virtual reality. Virtual Real. **11**, 261–273 (2007). https://doi.org/10.1007/s10055-007-0073-7
24. Yang, X., Liu, J., Lv, N., Xia, H.: A review of cable layout design and assembly simulation in virtual environments. Virtual Real. Intell. Hardw. **1**(6), 543–557 (2019). https://doi.org/10.1016/j.vrih.2019.11.001
25. Isohata, E., Takahashi, K., Ino, H.: 3D–2D interface CAD system for wiring harness. Fujikura Tech. Rev. **101**(2001), 61–65 (2001)
26. Čapek, R., Šůcha, P., Hanzálek, Z.: Scheduling of production with alternative process plans. In: Schwindt, C., Zimmermann, J. (eds.) Handbook on Project Management and Scheduling, vol. 2, pp. 1187–1204. LNCS. International Handbooks on Information Systems. Springer, Cham (2015). https://doi.org/10.1007/978-3-319-05915-0_23.
27. Nguyen, H.G., Kuhn, M., Franke, J.: Manufacturing automation for automotive wiring harnesses. Procedia CIRP **97**, 379–384 (2021). https://doi.org/10.1016/j.procir.2020.05.254
28. Benmamoun, Z., Fethallah, W., Ahlaqqach, M., Jebbor, I., Benmamoun, M., Elkhechafi, M.: Butterfly algorithm for sustainable lot size optimization. Sustainability **15**, 11761 (2023). https://doi.org/10.3390/su151511761
29. El Khalfi, A., Mahdou, N., Zahir, Y.: Strongly primary ideals in rings with zero divisors. Quaestiones Mathematicae, **44**(5), 569580 (2021). https://doi.org/10.2989/16073606. 2020. 1728416
30. Khalfi, A.E., Mahdou, N., Zahir, Y., Rings in which every nonzero weakly prime ideal is prime. São Paulo J. Math. Sci. **14**, 689–697 (2020). https://doi.org/10.1007/s40863-020-001 72-6
31. Gorostiza, C.Z., Hendrickson, C., Rehak, D.R.: Knowledge-Based Process Planning for Construction and Manufacturing. Elsevier, Amsterdam (1989)
32. Azeroual, M., et al.: Advanced energy management and frequency control of distributed microgrid using multi-agent systems. Int. J. Emerg. Electr. Power Syst. **23**(5), 755–766 (2022). https://doi.org/10.1515/ijeeps-2021-0298
33. Benmamoun, Z., Hachimi, H., Amine, A.: Comparison of inventory models for optimal working capital; case of aeronautics company. Int. J. Eng. **31**(4), 605–611 (2018)
34. Zheng, J., Zhong, J., Chen, M., He, K.: A reinforced hybrid genetic algorithm for the traveling salesman problem. Comput. Oper. Res. **157**, 106249 (2023). https://doi.org/10.1016/j.cor.2023.106249
35. Grefenstette, J.J., Gopal, R., Rosmaita, B.J., Van Gucht, D.: Genetic algorithms for the traveling salesman problem. In: Proceedings of the 1st International Conference on Genetic Algorithms, Pittsburgh, PA, USA, pp. 160–168 (1985)
36. Baha, A., Hasan, J.W., Mauro, O.: Assembly design semantic recognition using SolidWorks-API. Int. J. Mech. Eng. Robot. Res. **5**(4), 280–287 (2016). https://doi.org/10.18178/ijmerr.5.4.280-287
37. Benmamoun, Z., Hachimi, H., Amine, A.: Inventory management optimization using lean six-sigma Case of Spare parts Moroccan company, presented at the Proceedings of the International Conference on Industrial Engineering and Operations Management, 2017, pp. 1722–1730
38. Elkhechafi, M., Hachimi, H., Elkettani, Y.: A new hybrid firefly with genetic algorithm for global optimization. Int. J. Manag. Appl. Sci. **3**, 47–51 (2017)

39. Elkhechafi, M., Hachimi, H., Elkettani, Y.: A new hybrid cuckoo search and firefly optimization. Monte Carlo Methods Appl. **24**(1), 71–77 (2018)
40. Hong, J., Chiou, R.Y., Kwon, Y.J.: Information visualization of networked assembly robots. Int. J. Mech. Eng. Robot. Res. **4**(4), 331–335 (2015). https://doi.org/10.18178/ijmerr.4.4.331-335
41. Benmamoun, Z., Fethallah, W., Bouazza, S., Abdo, A.A., Serrou, D., Benchekroun, H.: A framework for sustainability evaluation and improvement of radiology service. J. Clean. Prod. **401**, 136796. https://doi.org/10.1016/j.jclepro.2023.136796

Use of Artificial Intelligence in Occupational Health and Safety in Construction Industry: A Proposed Framework for Saudi Arabia

Shabir Hussain Khahro[1(✉)] and Qasim Hussain Khahro[2]

[1] Department of Engineering Management, College of Engineering, Prince Sultan University, Riyadh, Saudi Arabia
shkhahro@psu.edu.sa

[2] Jamilus Research Center, Faculty of Civil Engineering and Built Environment, Universiti Tun Hussein Onn Malaysia, 86400 Parit Raja, Batu Pahat, Johor, Malaysia

Abstract. Occupational accidents have always been very important for all industries due to its significant impacts on projects and society. Whereas the case in construction industry is also similar due to higher number of occupational accidents recorded in various countries. In current dynamic technological era, conventional Health and Safety (H&S) practices are not sufficient thus new approaches are important to explore. It is observed from the detailed literature review that there are various existing tools designed by various researcher in different scenario but there is a lack to device an integrated approach. Therefore, this paper proposes an integrated approach to monitor the H&S at site. The proposed framework will help the decision makers to manage and monitor the health and safety practices at site in a real time environment.

Keywords: Construction Safety · Artificial Intelligence · Big Data · Worker Health · Future Technologies

1 Background

The construction industry has historically faced constant difficulties in terms of safety concerns. Globalization and the competitive environment of the 21st century provide numerous challenges for organizations. Safety compliance is one of the key factors in reducing workplace accidents and task complexity [1]. Poor construction safety has remained a major issue, because it results in accidents, injuries, or even fatalities for workers and road users [2]. The construction accidents, injuries, and fatalities result in significant losses and expenses annually. The incidence of occupational injuries and worker accidents in construction sites is significantly higher in comparison to other types of occupations. Despite significant advancements in occupational safety, effectively managing and controlling the risks associated with workplace activities remains a challenging one [3].

S.-H. Sheu (Ed.): ICIEA-EU 2024, LNBIP 507, pp. 49–59, 2024.
https://doi.org/10.1007/978-3-031-58113-7_5

Construction companies must ensure their employees continue to be productive without compromising their well-being. It is observed that the poor health and safety policies cost the worldwide construction industry billions of dollars every year [4]. This has inspired safety experts and researchers to find new procedures, tools, equipment, and laws to reduce the rate of accidents during the construction of any project. Construction safety management prevents accidents through planning and monitoring. Safety planning detects and evaluates jobsite hazards, including their likelihood and severity. Safety monitoring involves monitoring workers and their environment to prevent harmful acts and conditions [5]. Ensuring safety monitoring presents challenges due to the inherent difficulties of simultaneously visiting multiple regions and the presence of certain sites that are inaccessible or unsafe for inspection. The industry's lack of digitization and manual nature makes project management complicated and cumbersome. Insufficient digital skills and technology adoption in the construction business can lead to cost inefficiencies, project delays, poor quality, uninformed decision-making, and poor productivity, health, and safety. Recent years have shown that the construction sector must embrace digitalization and swiftly enhance technology capability to address the challenges for sustainable infrastructure [6].

Artificial intelligence (AI) has been making notable advancements in revolutionizing diverse industries, including the construction sector. Within the domain of Occupational Health and Safety (OHS), AI is being utilized to augment safety protocols, mitigate potential hazards, and better the overall administration of construction sites. The utilization of AI in the context of OHS within the construction sector is primarily focused on the reduction of accidents, enhancement of worker welfare, and establishment of safer work conditions. As the field of technology continues to advance, it is anticipated that these solutions will undergo further refinement, thereby enhancing the safety and productivity of the construction industry. AI possesses the capability to effectively analyze substantial quantities of data derived from many sources, including sensors, wearables, and historical records, in order to generate practical and implementable insights [7]. This facilitates the ability of management to make well-informed decisions aimed at enhancing safety rules and practices. AI algorithms possess the capability to examine both historical data and real-time inputs in order to forecast future dangers and risks that may arise within building sites. Through the analysis of accident and near-miss occurrences, artificial intelligence (AI) systems have the capability to recognize trends and subsequently propose preventive measures and safety regulations to effectively mitigate these risks.

AI-driven communication systems and applications have the capability to promptly notify employees on potential risks, alterations in weather conditions, or other unforeseen emergencies. This measure guarantees that employees are adequately informed and able to swiftly implement the appropriate precautions. AI is presently transforming various sectors with the aim of attaining heightened levels of safety and security [6]. [3] emphasized the necessity of implementing efficient control and safety protocols at the construction sites to mitigate or, ideally, eliminate the factors contributing to injuries and accidents. With the growing imperative to enhance health and safety management among workers in the construction industry, there is a growing demand for construction organizations to adopt and implement novel technologies in order to enhance workers' health and safety [4].

Recent developments in sensor technology and the Internet of Things (IoT) have made it possible to automate and conduct real-time monitoring of a variety of construction-related tasks (Rao et al., 2022). AI has the capability to enhance emergency response efforts through the analysis of worker locations, identification of potential risks, and determination of safe departure routes in times of emergency. This facilitates rapid and effective action of response teams. Researchers are getting advantages of AI to enhance the growth of the countries. However, organizations have realized that machines can improve construction safety and efficiency. Various studies have been conducted worldwide and proposed various solutions based on AI. [8] have worked on modeling, with a focus on artificial intelligence for predicting CO2 emissions for the purpose of mitigating the effects of climate change in Saudi Arabia. In China, [9] proposed a digital-twin based multi-information intelligent early warning and safety management platform in order to address the significant safety risks associated with tunnel construction. [10] aimed to test the viability of utilizing data from a wearable insole pressure system, specifically acceleration and foot plantar pressure distribution data, for the automated recognition of workers' actions associated with overexertion and the evaluation of the relevant ergonomic risk levels. [11] developed a smart garment for real-time ergonomic risk assessment, utilizing posture awareness to empower operators and deliver objective data to ergonomists.

AI has the potential to serve as a powerful instrument for organizations seeking to improve safety and health measures, while fostering an ideal atmosphere for overall growth and success. AI has the potential to significantly contribute to the improvement of Occupational Health and Safety (OHS) within the construction sector through a range of mechanisms:

1.1 Personal Protective Equipment (PPE)

The utilization of artificial intelligence (AI) in personal protective equipment (PPE) enables the monitoring of workers' physiological indicators, identification of weariness or indications of discomfort, and the provision of timely notifications in the event of emergencies. Sensor technology in PPE can monitor employees' health, exposure to dangerous elements, and proximity to danger zones. Using smartphone and wristwatch technology can increase worker awareness by gathering data on the workplace, recognizing environmental and health hazards. An example of this is the utilization of intelligent helmets that are equipped with cameras and sensors, enabling the analysis of worker behavior, identification of hazardous behaviors, and provision of immediate warnings in real-time [12] (Fig. 1).

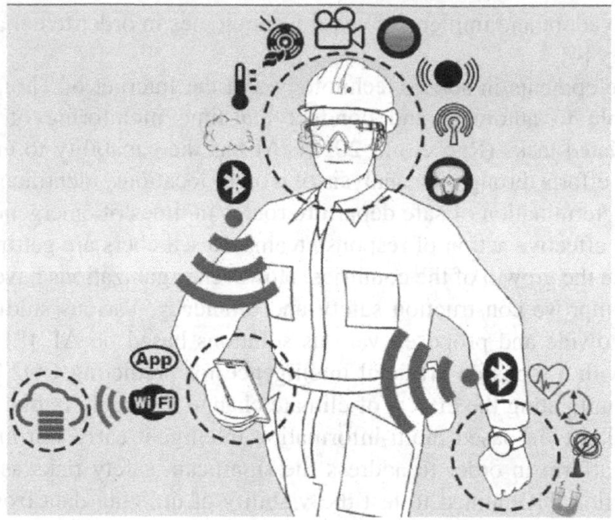

Fig. 1. PPE interconnected with smart technology [12]

The advent of wearable technology has broadened the scope of personal protective equipment (PPE) to encompass devices that facilitate communication between workers and project managers, offer contextual information to field personnel, and even detect environmental conditions, predict potential hazards, and broadcast wireless alarms.

1.2 Automated Safety Inspections

Drones that are outfitted with artificial intelligence (AI) have the capability to conduct automated safety inspections at building sites. The identification of structural faults, dangerous conditions, and compliance violations enables expedited reaction times and enhanced monitoring capabilities. Unmanned Aerial Vehicles (UAVs) are being used for the purpose of easily and effectively examining areas that are otherwise inaccessible, difficult to reach, or pose safety risks at building sites. This utilization of UAVs serves to improve the evaluation of safety conditions on construction projects [13]. In a study by [14] UAVs were found to be more effective than hand-held cameras due to their capacity to capture various angles and heights, ensuring complete jobsite coverage (Fig. 2).

Fig. 2. Unmanned aerial vehicles (UAV) images for safety monitoring at a construction site in Chile: (a) lack of guardrails; (b) worker without safety rope; and (c) lack of guardrails [13].

1.3 Video Analytics and Surveillance

The utilization of AI-powered video analysis enables the real-time monitoring of construction sites, facilitating the identification of potential safety violations, including instances where workers fail to adhere to proper safety protocols or engage in hazardous behaviors. Supervisors can receive automated alerts for prompt intervention [15] (Fig. 3).

Fig. 3. Examples of object detection at construction site [15].

1.4 Predictive Maintenance of Equipment

Through the analysis of equipment data, artificial intelligence (AI) has the capability to anticipate potential malfunctions or maintenance needs in machinery. The implementation of preventive measures serves to mitigate the occurrence of unforeseen malfunctions, hence reducing the likelihood of accidents and promoting the establishment of a safer working environment. In order to make maintenance work more proactive, predictive maintenance basically entails the process of anticipating a breakdown of the system that needs to be maintained by identifying early warning signals of failure [16].

1.5 Virtual Reality (VR) Training

Using virtual reality (VR) for safety training has reduced obstacles and enabled continuous training for the manpower working in construction industry. AI in conjunction with VR technology enables the creation of immersive safety training simulations, which can effectively enhance the training experiences of workers. The utilization of simulations enables workers to engage in the practice of hazardous jobs within a controlled environment, so enhancing their proficiency and self-assurance while mitigating the occurrence of real-world hazards. VR has been very helpful in monitoring the safety performance at construction projects [17].

Fig. 4. Scaffolding task safety training by VR [18]

1.6 Autonomous Vehicles and Equipment

Automatic construction vehicles and equipment driven by AI have the potential to mitigate the hazards inherent in machinery handled by humans. In addition, these vehicles have the capability to be outfitted with collision-avoidance systems, so augmenting their safety measures. The objective of this endeavor is to implement a combination of practical and advanced artificial intelligence techniques in vehicles, enabling them to exhibit behaviors that are comparable to or above human capabilities [19]. The Fig. 4 shows an unmanned excavator, developed by [20], possesses the capability to undertake jobs of high risk, including earthworks and demolition. Additionally, this system can be employed in hazardous recovery operations within disaster-stricken regions or areas deemed unsafe for human presence, such as high radiation zones within nuclear facilities (Fig. 5).

Fig. 5. Unmanned excavator [20]

1.7 Digital Twins for Safety Simulation

A digital twin is a virtual representation that accurately represents an artificial or physical system, incorporating strategically positioned sensors to collect diverse data pertaining to the system's operation [21]. For significant safety risk during tunnel building, [9] proposed a digital-twin-based multi-information intelligent early warning and safety management platform. The integration of artificial intelligence with digital twin technology enables the generation of virtual representations of construction sites. These models possess the capability to replicate a wide range of scenarios and evaluate potential safety hazards prior to the commencement of actual construction operations.

The above trends serve as evidence of the transformative impact of artificial intelligence (AI) on occupational health and safety within the construction sector. This technological advancement has the potential to significantly improve worker protection and mitigate potential hazards. With the ongoing advancement of technology, it is anticipated that these applications will undergo further development and achieve a higher level of integration within building site operations on a global scale.

2 Construction Industry in Saudi Arabia

One of the countries in the Middle East that is regarded as having one of the highest rates of economic growth is Saudi Arabia. The construction industry in Saudi Arabia holds the second position in terms of its contribution to the gross domestic product, with only the petroleum sector surpassing it [22]. However, this sector of the economy is home to some of the most dangerous workplaces in the entire country [3]. As the building industry in Saudi Arabia grows post-oil, it is crucial to study the occupational health and safety practices of construction workers [23].

Although many industrialized nations rely extensively on safety management, its significance in Saudi Arabia's construction industry is quite insignificant. In Saudi Arabia, construction quality has been a concern and inconsistently high cause of danger and accidents [2]. Despite the existence of this information, the overall level of construction safety in Saudi Arabia still needs improvements, which keeps keep the attention of the stakeholders [24].

To prevent construction accidents, safety managers must identify and assess workplace hazards, assessing their likelihood and severity by adopting the new trends being adopted for construction safety in the world.

3 Proposed Framework

This research proposed is an adaptive feedback system for detection of hazards, collisions, mishaps that learn from its own experiences as shown in Fig. 6. The system doesn't create decisions on traditional intelligent frameworks that are trained of very old data sets rather it relies on real time inputs and feedback after the alert action. The framework is divided into three levels (1) Inputs (2) Intelligent Data Fusion and (3) AI decision making and reporting.

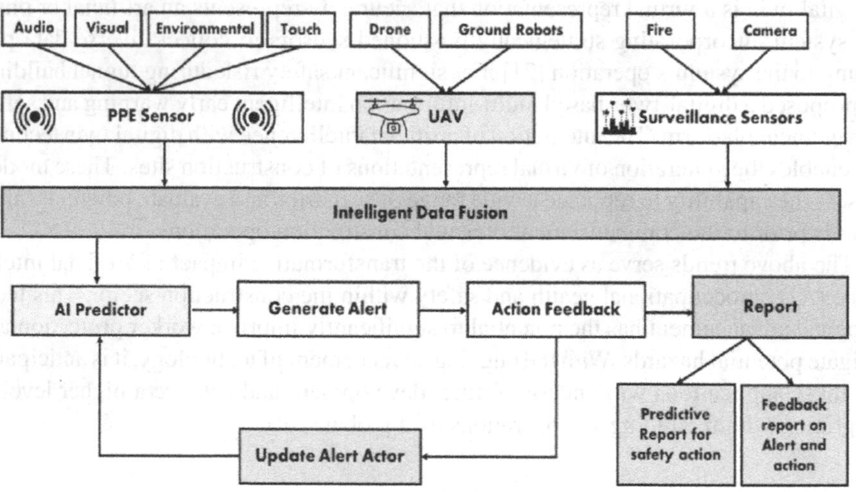

Fig. 6. Showing the proposed framework for intelligent safety hazard detection and prevention.

3.1 Input Module

Although the system heavily relies on inputs, but this framework divides the inputs into three general categories. (1) PPE Sensor, (2) Unmanned Autonomous Vehicle (UAV) and (3) Surveillance sensors. The first input of the system relies on sensors such as visual sensors that are usually surveillance cameras, audio sensors to sense any unknown or distress alerts, the touch sensor can detect the intensity of collision and take proper actions. Environmental sensors such as smoke, vibration and other sensors can also assist in detection of any possible hazards. This research suggests having some of these sensors as a part of PPE such as camera showing where the construction worker is, and if there is potential hazard in front of him then alert the worker and authorities. Second input is from building installed sensors to find any potential hazard. These sensors can vary from smoke sensors to humidity sensors or even fire sensors. Third and live input is from ground robots and UAV allowing them to reach places where humans cannot go easily or can be used in speeding up the recovery process.

3.2 Intelligent Data Fusion (IDF) Module

There are multiple inputs in the proposed framework but if these inputs are not combined and are not linked plus stitched together then it cannot make real sense or can have in figuring out the intensity of issue. IDF module will use an intelligent approach to combine all the data from these inputs and make it meaningful data.

3.3 AI Decision Making and Reporting Module

Once the data is fused together the data is forwarded to AI predictor module that takes the decision to generate an alert of not, secondly it will also decide severity of data, thirdly it will decide which safety parties should be addressed. Then AI predictor generates an alert. Once the Alert is generated and actions are taken the system asks for feedback from all safety parties and adapt two-way approach. The first one is feedback to the main system to help AI predictor in future if similar or near similar situation arise. The second one is the report generation for the management both internal and government related to situation. The report module also performs an extra step, that is generate case studies for training and awareness sessions where it takes care of privacy issues, approval issues whether such incidents can be shared and what can be shared. This module will help in assisting trainees and students to stay in touch with up-to-date knowledge and will also bridge the gap between industry and academia. The future director can also result in an online asynchronous course that can be offered without the assistance of any trainer and can help the global workforce.

4 Conclusion

Each year, enormous losses and costs are incurred as a result of construction-related accidents, injuries, and fatalities. Compared to other occupations, construction sites have a much greater frequency of worker accidents and occupational injuries. Despite substantial improvements in occupational safety, it is still difficult to properly manage and control the hazards related to working operations.

The first one is feedback to the primary system to aid AI predictor in the future should a circumstance similar to or nearly identical to the current one emerge. The creation of reports for management, both internal and governmental, is the second one. The report module also completes an additional task, which is to provide case studies for training and awareness sessions where it addresses privacy concerns, permission concerns regarding whether such instances may be shared, and approval concerns regarding what can be shared. Through the utilization of artificial intelligence (AI) technology, construction organizations experience the ability to establish work environments that prioritize safety, resulting in a reduction of accidents, mitigation of health risks, boost towards production, and development of a safety-oriented culture among workers. These innovations not only contribute to the safeguarding of the welfare of construction employees but also result in financial savings and enhanced project outcomes.

Acknowledgments. The authors are thankful to Prince Sultan University, Saudi Arabia for supporting this conference attendance and expert support.

References

1. Hofmann, D.A., Burke, M.J., Zohar, D.: 100 years of occupational safety research: from basic protections and work analysis to a multilevel view of workplace safety and risk. J. Appl. Psychol. **102**(3), 375–388 (2017). https://doi.org/10.1037/apl0000114
2. Al-Otaibi, A., Kineber, A.F.: Identifying and assessing health and safety program implementation barriers in the construction industry: a case of Saudi Arabia. Appl. Sci. (Switzerland) **13**(4) (2023). https://doi.org/10.3390/app13042630
3. Abukhashabah, E., Summan, A., Balkhyour, M.: Occupational accidents and injuries in construction industry in Jeddah city. Saudi J. Biol. Sci. **27**(8), 1993–1998 (2020). https://doi.org/10.1016/j.sjbs.2020.06.033
4. Ibrahim, K., Simpeh, F., Adebowale, O.J.: Benefits and challenges of wearable safety devices in the construction sector. Smart and Sustainable Built Environment (2023). https://doi.org/10.1108/SASBE-12-2022-0266
5. Perlman, A., Sacks, R., Barak, R.: Hazard recognition and risk perception in construction. Saf. Sci. **64**, 13–21 (2014). https://doi.org/10.1016/j.ssci.2013.11.019
6. Abioye, S.O., et al.: Artificial intelligence in the construction industry: a review of present status, opportunities and future challenges. J. Build. Eng. **44**. Elsevier Ltd (2021). https://doi.org/10.1016/j.jobe.2021.103299
7. Ahn, C.R., Lee, S., Sun, C., Jebelli, H., Yang, K., Choi, B.: Wearable Sensing Technology Applications in Construction Safety and Health. J. Constr. Eng. Manag. **145**(11) (2019). https://doi.org/10.1061/(asce)co.1943-7862.0001708
8. Alamri, S., Khan, S., Town, I.: Artificial intelligence based modelling for predicting CO 2 emission for climate change mitigation in Saudi Arabia (IJACSA). Int. J. Adv. Comput. Sci. Appl. **14**(4), 182–189 (2023). www.ijacsa.thesai.org
9. Ye, Z., et al.: A digital twin approach for tunnel construction safety early warning and management. Comput. Ind. **144** (2023). https://doi.org/10.1016/j.compind.2022.103783
10. Antwi-Afari, M.F., Li, H., Umer, W., Yu, Y., Xing, X.: Construction activity recognition and ergonomic risk assessment using a wearable insole pressure system. J. Constr. Eng. Manag. **146**(7) (2020). https://doi.org/10.1061/(asce)co.1943-7862.0001849
11. Cerqueira, S.M., Da Silva, A.F., Santos, C.P.: Smart vest for real-time postural biofeedback and ergonomic risk assessment. IEEE Access **8**, 107583–107592 (2020). https://doi.org/10.1109/ACCESS.2020.3000673
12. Adjiski, V., Despodov, Z., Mirakovski, D., Serafimovski, D.: System architecture to bring smart personal protective equipment wearables and sensors to transform safety at work in the underground mining industry. Rudarsko Geolosko Naftni Zbornik **34**(1), 37–44 (2019). https://doi.org/10.17794/rgn.2019.1.4

13. Martinez, J.G., Gheisari, M., Alarcón, L.F.: UAV integration in current construction safety planning and monitoring processes: case study of a high-rise building construction project in Chile. J. Manag. Eng. **36**(3) (2020). https://doi.org/10.1061/(asce)me.1943-5479.0000761
14. Tuttas, S., Braun, A., Borrmann, A., Stilla, U.: Acquisition and consecutive registration of photogrammetric point clouds for construction progress monitoring using a 4D BIM. Photogrammetrie Fernerkundung Geoinformation **85**(1), 3–15 (2017). https://doi.org/10.1007/s41064-016-0002-z
15. Kim, J.: Visual analytics for operation-level construction monitoring and documentation: state-of-the-art technologies, research challenges, and future directions. Front. Built Environ. **6**. Frontiers Media S.A. (2020). https://doi.org/10.3389/fbuil.2020.575738
16. Selcuk, S.: Predictive maintenance, its implementation and latest trends. Proc. Inst. Mech. Eng. B J. Eng. Manuf. **231**(9), 1670–1679 (2017). https://doi.org/10.1177/0954405415601640
17. Rokooei, S., Shojaei, A., Alvanchi, A., Azad, R., Didehvar, N.: Virtual reality application for construction safety training. Saf. Sci. **157** (2023). https://doi.org/10.1016/j.ssci.2022.105925
18. Yoo, J.W., Park, J.S., Park, H.J.: Understanding VR-based construction safety training effectiveness: the role of telepresence, risk perception, and training satisfaction. Appl. Sci. (Switzerland) **13**(2) (2023). https://doi.org/10.3390/app13021135
19. Li, J., Cheng, H., Guo, H., Qiu, S.: Survey on artificial intelligence for vehicles. Automot. Innov. **1**(1), 2–14 (2018). https://doi.org/10.1007/s42154-018-0009-9
20. J. Lee, B. Kim, D. Sun, C. Han, and Y. Ahn, "Development of unmanned excavator vehicle system for performing dangerous construction work," Sensors (Switzerland), vol. 19, no. 22, Nov. 2019, doi: https://doi.org/10.3390/s19224853
21. Piromalis, D., Kantaros, A.: Digital twins in the automotive industry: the road toward physical-digital convergence. Appl. Syst. Innov. **5**(4). MDPI (2022). https://doi.org/10.3390/asi5040065
22. SCA., Statistical Report of Contracting Sector 2018, 2018. Accessed 07 Sep 2023
23. Alsulami, H., Serbaya, S.H., Rizwan, A., Saleem, M., Maleh, Y., Alamgir, Z.: Impact of emotional intelligence on the stress and safety of construction workers' in Saudi Arabia. Eng. Constr. Archit. Manag. **30**(4), 1365–1378 (2023). https://doi.org/10.1108/ECAM-06-2021-0481
24. Al Haadir, S., Panuwatwanich, K.: Critical success factors for safety program implementation among construction companies in Saudi Arabia. Procedia Eng. 148–155 (2011). https://doi.org/10.1016/j.proeng.2011.07.017

Path Score Based Approach for Deriving Machine Sequence in Multistage Process

Sugyeong Lee[1,2] and Dong-Hee Lee[3](✉)

[1] Department of Semiconductor and Display Engineering, Sungkyunkwan University, 2066 Seobu-ro, Suwon, Korea
sugyeong@g.skku.edu
[2] Mobile Display Division, Samsung Display Co., Ltd., 181 Samsung-ro, Asan, Korea
[3] Department of Industrial Engineering, Sungkyunkwan University, 2066 Seobu-ro, Suwon, Korea
dhee@skku.edu

Abstract. Path selection in multistage manufacturing process has a considerable impact on product quality because uniform product quality cannot be ensured across all equipment. Accordingly, we present a method for choosing the best path while taking product quality into account. In this paper, we firstly introduce a quality index and categorize product grades. Building upon the criteria, we also propose an approach based on path scores to identify machine sequence paths that can improve product quality. To validate its efficacy, we apply the approach to the industry-like data, successfully pinpointing both the most and least favorable paths.

Keywords: Multistage manufacturing process (MMP) · Path selection · Product quality · Quality engineering · Semiconductor · Display

1 Introduction

The silicon engineering including semiconductor wafer fabrication process and display backplane manufacturing process is composed of hundreds of steps, involving multiple machines in each single step. While it would be ideal for all equipment at each process stage to produce outcomes with the same quality, it is practically unattainable. Consequently, finding the optimal path is essential to maximize quality enhancement. However, studies on path selection and wafer assignment so far have primarily focused on the productivity [6].

As quality standards are escalating and technology is rapidly advancing in the industry, the quality has become a significant concern. So far, when it comes to the quality, yield has been the paramount performance indicator in the semiconductor industry [10]. Along this belief, studies considering both quality and productivity perspective have been recently conducted. Lee suggested a method to derive golden path to maximize yield [5]. Still, this study also has a limitation in that it solely considers yield as the quality metric.

S.-H. Sheu (Ed.): ICIEA-EU 2024, LNBIP 507, pp. 60–70, 2024.
https://doi.org/10.1007/978-3-031-58113-7_6

Meanwhile, processes in the silicon engineering have reached a level of maturity and ensured a yield above a certain level in most cases. As a matter of fact, it has been more than four decades since the initiation of semiconductor production in Korea, and nearly 3 decades have passed since display production began as well [1]. In this context, it is highly desirable to subdivide the product quality into several classes to improve quality. To maximize quality, a prior discussion on how to define quality grades is necessary.

Therefore, this paper outlines the logic for sorting products at display panel level and distinguishing products at the mother glass level that stand out from others. We have taken a step further by introducing an additional metric called 'good panel ratio' to define quality grades. Using both yield and good panel ratio as criteria, we have established a new set of standards for defining quality grades.

This study also aims to identify the best and worst path for the quality perspective. We have developed an algorithm to derive the best and worst path for quality enhancement by formulating path score. Then the algorithm is applied to industry-like data to validate its effectiveness.

First, we review previous studies related to path selection conducted so far in Sect. 2. In Sect. 3, we present the methodology to differentiate the quality and obtain paths for maximizing quality. Then we implement the method to the practical dataset which is designed to reflect some part of display backplane process, and confirm whether the proposed method is performed admirably in Sect. 4. Finally, we conclude the paper by discussing the results and suggesting future improvement of the proposed approach.

2 Related Works

Kickstarting from Wein's research that indicated scheduling has a significant impact on cycle time [11], scheduling on semiconductor has evolved diversely. Now it is obvious that lot scheduling and path selection directly affect the productivity, and outcomes may vary depending on whether time-based, utilization-based, or delivery-based criteria are prioritized even if they are all categorized as productivity factors [6].

This means semiconductor manufacturing system is a huge complex system that is difficult to analyze and optimize. So, Senties et al. suggested several performance criteria and interpreted scheduling problem to multi-objective optimization of combinations of criteria pairs including machine utilization, cycle time, and waiting time which are all related to productivity [9]. Lee et al. also modelled multi-objective scheduling considering unit profit, tardiness cost and WIP inventory [7].

However, there are few studies that consider the quality as a performance indicator. Fortunately, many studies have started to be conducted for yield prediction to minimize defective wafers recently [3,4], yet hardly related to path selection.

To overcome this limitation, Lee et al. have dedicated significant effort towards developing a method of wafer assignment to maximize yield as well as productivity simultaneously [5,6]. While building on prior research, we introduce the novelty by incorporating a comprehensive quality perspective that accounts for both quality level and yield, as opposed to solely emphasizing yield for quality measure as in previous studies.

3 Proposed Method

3.1 Industrial Background

Display manufacturing has 2 parts, array and cell process (i.e. front-end process) and module process (i.e. back-end process) [8]. Front-end process deals with display backplane mother glasses and back-end process deals with panels. Figure 1 shows the structure of the mother glass and panels. In back-end process, panels are cut from mother glass and then several inspection and rework stages are conducted to each panel.

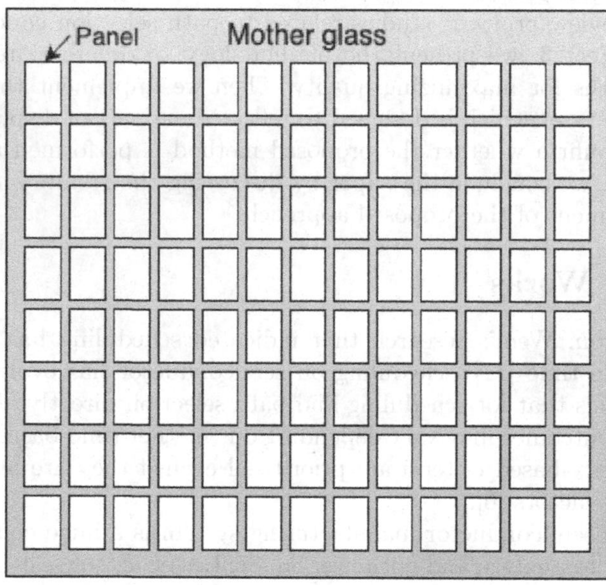

Fig. 1. General display glass and panel structure

Mainly, yield is calculated at the panel level in back-end process after various inspection and rework stages. As a result, if there does not exist any defect on panel or hardly exists so that the product is superior to cosmetic specification, the panel is classified as good panel (i.e. level 1). If there exist mimor defects on panel so in-depth test should be conducted, but still the product meets qualification, the panel is classified as level 2. If there exist major or combined minor defects on panel but they can be repaired after various rework stages, the panel is classified as level 3. Otherwise, the panel is scrapped because there exist irrepairable major defects on panel that are not qualified to customer requirements.

Only panels to which the level is assigned can be delivered to customer. Thus, yield is calculated as the ratio of the number of graded panels to the total number of panels. Also, the ratio of the number of panels with level 1 to the total number of panels can be called good panel ratio (hereinafter referred to as GPR).

Note that yield is calculated on a panel basis, yet a mother glass is the primary unit of production. Hence, it is crucial to manage yield and GPR on a glass level in order to enhance manufacturing process. Within this framework, we define a comprehensive quality grading logic of mother glasses with both yield and GPR as follows. If both yield and GPR exceed some thresholds, the glass-level quality grade is defined as S. If yield exceed the threshold but GPR does not, the quality grade is defined as A. Otherwise, the quality grade is defined as B.

In manufacturing industry where technology has matured sufficiently, yield is already somewhat guaranteed. Therefore, improving the quality level through metrics like GPR becomes crucial, rather than solely relying on yield. We can accordingly define the comprehensive quality grade based on both yield and GPR data. Notably, these metric are being managed to achieve stretch goal, so the criteria that distinguish grade S from grade A and B can be expressed as combination of challenging target of yield and GPR.

3.2 Proposed Method

The idea of the proposed method is to identify the best and worst path among all feasible paths. The best path is the path that maximizes the quality. The worst path is the path that minimizes the quality and meets cosmetic specification at a minimum level. Although the product from the worst path can be still deliverable to the customer, it is better to refrain from doing so.

A simple diagram of procedure is drawn in Fig. 2. First, we calculate average grade S glass ratio from the entire dataset. Then, we search combinations of two adjacent processes which are above average grade S glass ratio to narrow down candidates for best path. Finally, we compute path score by adding path for each combination and find the best path.

Detailed procedure is described in Algorithm 1. After calculation of net grade S glass ratio, we examine interaction between two consecutive process stages if the grade S glass ratio of the interaction is above the average which has been calculated in Step 1. If true, we append the pair of process stages to the interaction dataset. This allows us to refine the candidates for the best path.

Subsequently, starting from the first process stage and adding machine of the following process stage within the interaction dataset, we compute path score to derive the best path. At this step, we utilize average yield and GPR per path to design the score. In case average yield of the path incorporating following process stage is higher than the previous value, then add 1 to the existing score. IF average GPR is higher than the previous one, then also add 1 to the existing

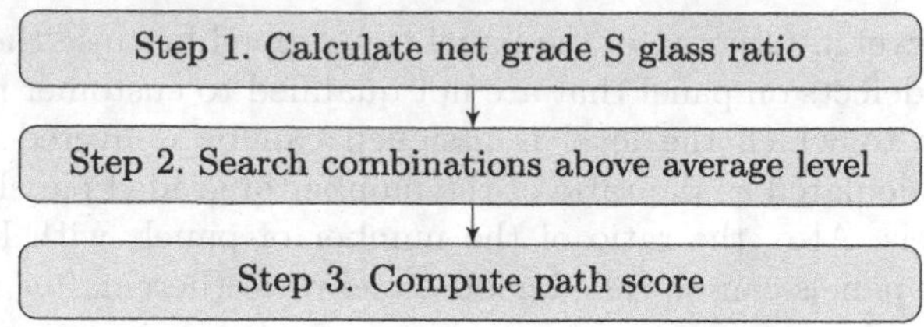

Fig. 2. Procedure of the proposed method

score. However, if average GPR is lower than the previous one, then subtract 1 from the score because we do not desire a decline in quality after a process has been carried out.

The best path is the path with the highest score. For the worst path case, it can be easily derived by reversing the sign of the algorithm.

Algorithm 1. Deriving the best path with quality indices

Require: dataset D, quality level criteria C, total number of process stage N
Calculate net grade S glass ratio R_{total} with C
for process stage $i = 1$ to $N - 1$ **do**
    Calculate grade S ratio from $P_{i,j}$ to $P_{i+1,k}$ for all j, k in D
    Denote the ratio as $R_{i,j\_i+1,k}$
    if $R_{i,j\_i+1,k} \geq R_{total}$ **then**
        Append path pair $\langle P_{i,j}, P_{i+1,k} \rangle$ to $D_{interaction}$
    end if
end for
Start with process stage 1 (i.e. $path = P_{1,n}$)
Define path score *score* and start with $score = 0$
while process stage $i < N$ and $\langle P_{i,j}, P_{i+1,k} \rangle \in D_{interaction}$ **do**
    if average yield $Y_{\langle path, P_{i+1,k} \rangle} \geq Y_{path}$ **then**
        $score += 1$
    end if
    if average GPR $GPR_{\langle path, P_{i+1,k} \rangle} > GPR_{path}$ **then**
        $score += 1$
    else if $GPR_{\langle path, P_{i+1,k} \rangle} < GPR_{path}$ **then**
        $score -= 1$
    end if
    Append path $P_{i+1,k}$ to existing *path*
    return *path*, *score*
end while
Best path = argmax *score*

---

4 Experimental Result

4.1 Dataset

Front-end process in display manufacturing is based on multi-layer structure devices, which closely resembles the manufacturing process of semiconductors. Much like semiconductor manufacturing, the whole display manufacturing process follows a large sequential set of similar subsets of processes. Furthermore, the sequences of processes within each subset are unidirectional.

Our industrial dataset is hypothetically generated to be a reflection of the practical paths taken during the manufacturing process of general display products. This dataset particularly captures a single subset of the entire process set, which comprises essential processes that are deeply related to the quality of the end product.

The dataset consists of the historical path of 2,650 mother glasses and corresponding yield and GPR information of each glass. Yield and GPR information follows Poisson distribution yield prediction model which is commonly used in the industry [2] and also been confirmed to fit well with the real-world industrial data.

Our dataset is composed of 5 process stages and each involving a different number of equipment in the following order respectively: 5, 7, 7, 4, and 8. These 5 process stages can represent some part of important processes in the display backplane manufacturing. Illustrative data sample is given in Table 1.

Table 1. Illustrative example of dataset

Glass No.	P1	P2	P3	P4	P5	Yield	GPR
1	$P_{1,1}$	$P_{2,1}$	$P_{3,1}$	$P_{4,2}$	$P_{5,1}$	***	***
2	$P_{1,1}$	$P_{2,1}$	$P_{3,1}$	$P_{4,2}$	$P_{5,1}$	***	***
3	$P_{1,1}$	$P_{2,1}$	$P_{3,1}$	$P_{4,2}$	$P_{5,1}$	***	***
4	$P_{1,1}$	$P_{2,1}$	$P_{3,1}$	$P_{4,2}$	$P_{5,2}$	***	***
...							
2650	$P_{1,5}$	$P_{2,7}$	$P_{3,7}$	$P_{4,1}$	$P_{5,8}$	***	***

As shown in Sect. 3.1, the quality index is determined by the yield and GPR of the glass. In the paper, according to challenging target of the internal quality criteria which is commonly relatable for our business, the quality level is defined as grade S if yield and GPR exceeds some specific challenging thresholds. Segments with the yield of over a relaxed threshold and outside of grade S are classified as grade A, while the rest are categorized as grade B.

Visualization figure of distribution of yield and GPR with grading guideline is in Fig. 3. Grade S glasses are represented as red dashed lines and dividing line between grade A and B is represented as blue dotted line. Also, area with grade S is hatched with blue color.

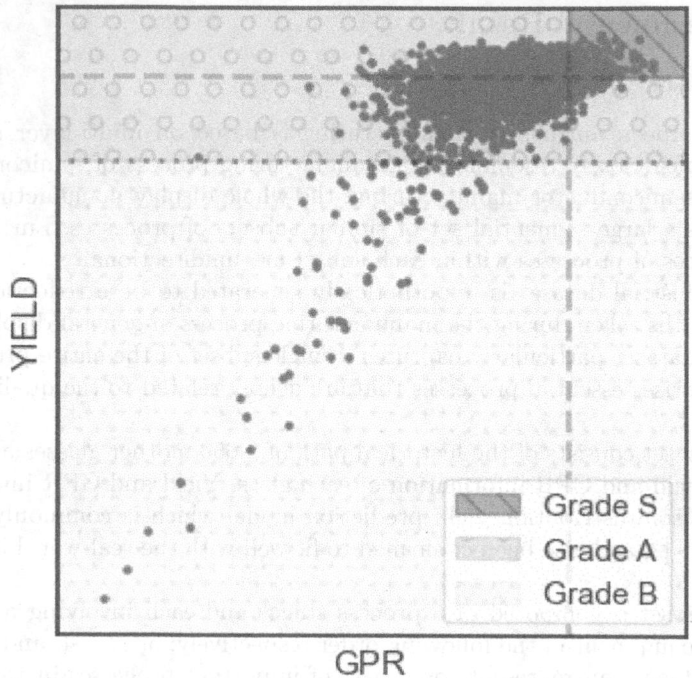

Fig. 3. Distribution of yield and GPR (Color figure online)

4.2 Result

Verification. In order to verify the effectiveness of the proposed algorithm, we apply our method to the dataset described in Sect. 4.1.

Calculating average grade S glass ratio from the entire dataset as described in Step 1 of Fig. 2, we get 0.0626 as a result for grade S glass ratio. This means that grade S glass ratio of the entire dataset is 6.26% and the number is 166 of 2650 glasses. To find the worst case, grade B glass ratio is 2.83% and the corresponding number is 75 out of the total.

After Step 2, the result as yield and GPR distribution is shown in Fig. 4. Fig. 4(a) shows the result of the best path candidates and Fig. 4(b) shows the worst path candidates. Resulting from Step 2, average glass S ratio of the best path candidates is 0.253 (43/170 glasses), and average glass B ratio of the worst path candidates is 0.963 (26/27 glasses), which is much higher than the average glass S/B ratio of the entire dataset.

Finally after Step 3, we get the best and worst path. The best path is $P_{1,5} - P_{2,3} - P_{3,1} - P_{4,1} - P_{5,2}$ and the glass S ratio is 66.7% (6/9 glasses). The worst path is $P_{1,3} - P_{2,6} - P_{3,1} - P_{4,2} - P_{5,4}$ and the glass B ratio is 100% (19/19 glasses). Fig. 5(a) shows the best path and Fig. 4(b) shows the worst path. This indicates that the proposed method successfully finds the best and worst path.

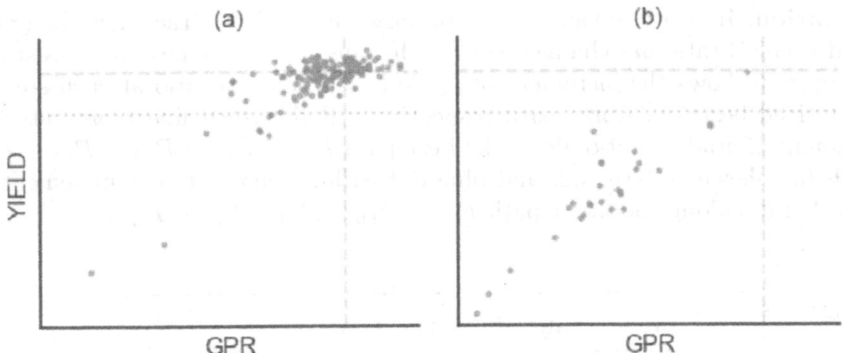

Fig. 4. Interim result - Yield and GPR after Step 2

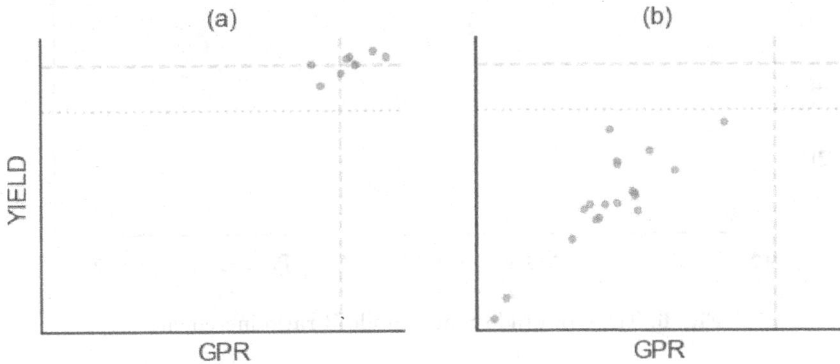

Fig. 5. Final Result - Yield and GPR after Step 3

As the result of the proposed method, the increment of grade S and grade B ratio followed by each step of Algorithm 1 is summarized in Table 2. In Table 2, (a) represents the increment of grade S rate after each step of the best path searching algorithm, and (b) represents the increment of grade B rate when searching the worst path. Note that it is highly advisable to avoid following the worst path because the outcome of this path worsens grade B rate to 100%.

Table 2. Result of the proposed method

	Grade	Initial value	After Step 2		After Step 3	
		Ratio	Ratio	Increment	Ratio	Increment
(a)	S	6.26%	25.3%	+19%p	66.7%	+41.4%p
(b)	B	2.63%	96.3%	+93.7%p	100%	+3.7%p

Validation. In order to validate the proposed method, we trace how the grade S and grade B ratio has changed with each transition to the next process stage.

Figure 6 shows the increment of grade S and grade B ratio at each step of the derived best and worst path respectively. Red dashed line represents the increment of grade S ratio along the best path $P_{1,5} - P_{2,3} - P_{3,1} - P_{4,1} - P_{5,2}$ which has shown in Sect. 4.2, and blue dotted line represents the increment of grade B ratio along the worst path $P_{1,3} - P_{2,6} - P_{3,1} - P_{4,2} - P_{5,4}$.

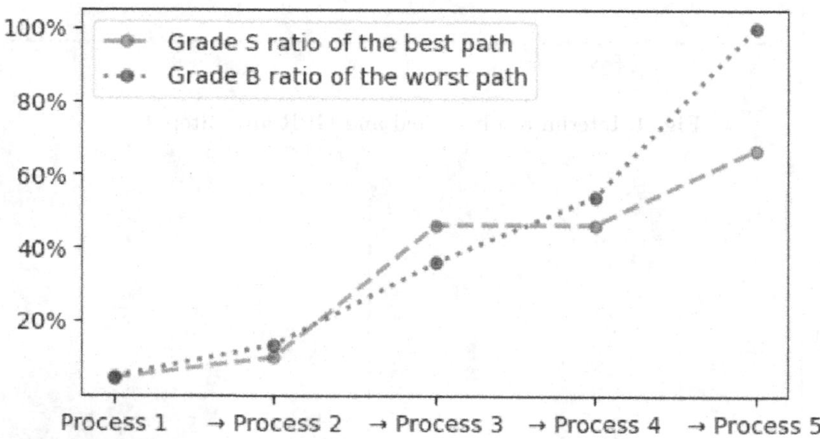

Fig. 6. Trace of grade S and grade B ratio increment

Table 3 presents the grade S and grade B ratio of each individual process stage. Within each process stage, the portion with the highest grade S and grade B ratio is highlighted in bold. The concept of the best path naturally emerges by merging machines associated with the highest ratios at every process stage. Identifying this particular machine sequence as the intuitive path, we proceed to compare it with the optimal path derived from our methodology.

For the grade B ratio, the intuitive path and the derived path by our method are identical. However, the derived path for the grade S ratio is significantly better than the intuitive path. In Table 4, the intuitive path is denoted by machine sequence (a), encompassing the combination of machines with the highest grade S ratio at each step. Meanwhile, machine sequence (b) represents the most superior path achieved through our proposed method. The grade S ratio for (a) stands at 33%, whereas (b) boasts a grade S ratio of 67%. Although the grade S ratio of (a) is commendable, it is unquestionable that the grade S ratio of (b) is twice as favorable. This disparity indicates that the path generated by our method is significantly more effective than the intuitive path.

Table 3. Grade S and grade B ratio of each process

(a) Process 1

#	S ratio	B ratio
$P_{1,1}$	**9.6%**	1.9%
$P_{1,2}$	7.1%	2.0%
$P_{1,3}$	6.3%	**4.7%**
$P_{1,4}$	6.2%	2.8%
$P_{1,5}$	4.2%	1.4%

(b) Process 2

#	S ratio	B ratio
$P_{2,1}$	**11.5%**	4.9%
$P_{2,2}$	10.0%	3.1%
$P_{2,3}$	7.1%	2.1%
$P_{2.4}$	6.4%	0.6%
$P_{2,5}$	4.8%	1.0%
$P_{2,6}$	4.6%	**5.2%**
$P_{2,7}$	4.3%	0.6%

(c) Process 3

#	S ratio	B ratio
$P_{3,1}$	**8.8%**	**6.7%**
$P_{3,2}$	7.7%	0.4%
$P_{3,3}$	7.4%	3.6%
$P_{3.4}$	6.0%	3.0%
$P_{3,5}$	4.6%	1.9%
$P_{3,6}$	4.4%	1.1%
$P_{3,7}$	3.3%	1.6%

(d) Process 4

#	S ratio	B ratio
$P_{4,1}$	**6.7%**	1.3%
$P_{4,2}$	6.5%	**5.8%**
$P_{4,3}$	6.4%	0.4%
$P_{4.4}$	3.5%	2.8%

(e) Process 5

#	S ratio	B ratio
$P_{5,1}$	**10.1%**	1.7%
$P_{5,2}$	7.2%	4.5%
$P_{5,3}$	6.3%	1.3%
$P_{5.4}$	6.2%	**10.9%**
$P_{5,5}$	6.0%	2.5%
$P_{5,6}$	5.5%	0.0%
$P_{5,7}$	5.1%	1.8%
$P_{5,8}$	4.4%	0.2%

Table 4. Comparison between the intuitive and derived path

	Machine sequence	Grade S ratio
(a)	$P_{1,1} - P_{2,1} - P_{3,1} - P_{4,1} - P_{5,1}$	33%
(b)	$P_{1,5} - P_{2,3} - P_{3,1} - P_{4,1} - P_{5,2}$	67%

5 Conclusion

In this paper, we suggest a method to find the path to enhance quality level of display backplane glass manufacturing. The proposed approach exploits yield and GPR information based on individual panels to interpret a glass-level quality and establish a new quality grading indicator. Next, based on this index, we have defined a path score for each machine sequence. Subsequently, we have developed an algorithm to find the best and worst path by maximizing this score.

It has been demonstrated to discover the best and worst path from the industry-like dataset, which can enhance overall display product quality. We expect that our proposed method can be comprehensively applied to any manufacturing process with multiple steps such as semiconductor manufacturing process.

Meanwhile, there is potential to expand the proposed method to increase its level of sophistication. First, the method lacks consideration for productivity factors. Another example of further investigation could be to extend the proposed method to model the relationship between path dataset and in-fab data such as delay time between processes. Also, it would be practicable to consider the case if machine degradation occurs during the manufacturing process.

References

1. Choung, J.Y., Hwang, H.R., Choi, J.H., Rim, M.H.: Transition of latecomer firms from technology users to technology generators: Korean semiconductor firms. World Dev. **28**(5), 969–982 (2000). https://doi.org/10.1016/S0305-750X(99)00161-8
2. Cunningham, J.: The use and evaluation of yield models in integrated circuit manufacturing. IEEE Trans. Semicond. Manuf. **3**(2), 60–71 (1990). https://doi.org/10.1109/66.53188
3. Jang, S.J., Kim, J.S., Kim, T.W., Lee, H.J., Ko, S.: A wafer map yield prediction based on machine learning for productivity enhancement. IEEE Trans. Semicond. Manuf. **32**(4), 400–407 (2019). https://doi.org/10.1109/TSM.2019.2945482
4. Jiang, D., Lin, W., Raghavan, N.: A novel framework for semiconductor manufacturing final test yield classification using machine learning techniques. IEEE Access **8**, 197885–197895 (2020). https://doi.org/10.1109/ACCESS.2020.3034680
5. Lee, C.H., Lee, D.H., Bae, Y.M., Choi, S.H., Kim, K.H., Kim, K.J.: Approach to derive golden paths based on machine sequence patterns in multistage manufacturing process. J. Intell. Manuf. **33**, 1–17 (2022). https://doi.org/10.1007/s10845-020-01654-2
6. Lee, D.H., Lee, C.H., Choi, S.H., Kim, K.J.: A method for wafer assignment in semiconductor wafer fabrication considering both quality and productivity perspectives. J. Manuf. Syst. **52**, 23–31 (2019). https://doi.org/10.1016/j.jmsy.2019.05.006
7. Lee, Y., Jiang, Z., Liu, H.: Multiple-objective scheduling and real-time dispatching for the semiconductor manufacturing system. Comput. Oper. Res. **36**(3), 866–884 (2009). https://doi.org/10.1016/j.cor.2007.11.006
8. Park, Y., Shintaku, J., Hong, P.: Effective supply chain configurations in the business context of china. MMRC Discuss. Study 1–19 (2012). https://www.researchgate.net/publication/266231755
9. Senties, O.B., Azzaro-Pantel, C., Pibouleau, L., Domenech, S.: Multiobjective scheduling for semiconductor manufacturing plants. Comput. Chemi. Eng. **34**(4), 555–566 (2010). https://doi.org/10.1016/j.compchemeng.2010.01.010
10. Tirkel, I., Rabinowitz, G., Price, D., Sutherland, D.: Wafer fabrication yield learning and cost analysis based on in-line inspection. Int. J. Prod. Res. **54**(12), 3578–3590 (2016). https://doi.org/10.1080/00207543.2015.1106609
11. Wein, L.: Scheduling semiconductor wafer fabrication. IEEE Trans. Semicond. Manuf. **1**(3), 115–130 (1988). https://doi.org/10.1109/66.4384

A Traffic Signal Timing Model with Consideration on Road Configurations, Phase Planning, and Conflict Points at a Signalized T-Intersection Road

Ma. Dominique M. Soriano and Ronaldo V. Polancos(✉)

De La Salle University, Manila, Philippines
{ma_dominique_soriano,ronaldo.polancos}@dlsu.edu.ph

Abstract. Continuous increase in demand for land transport calls for motorists' heightened usage of roads. This interconnectivity of road infrastructures features the various types of road intersections seen worldwide. When constructed, roads may be made functional with or without traffic lights. This study will feature a traffic management system for a signalized t-intersection to mitigate vehicular traffic and road congestion. A simple single-objective optimization model was developed to minimize the traffic system's overall cycle time and determine the green, red, and amber timing using the model and a traffic simulation to observe vehicle density, delay, and volume trends. From an unsignalized base model, three (3) signalized simulations were conducted with varying road configurations and phase plans to address conflict points in a t-intersection. A significantly efficient trend in conflict point reduction and phase planning improvement was achieved, meeting the objective of minimizing the cycle time.

Keywords: Optimization Modeling · Traffic System Management · Signalized T-Intersection · Traffic Simulation · Traffic Signal Timing

1 Introduction

A road network is the interconnectivity of the road infrastructures. Roads in all parts of the world are connected for maximum accessibility. Various types of road intersections characterize these. An intersection with timed traffic signals is called a signalized intersection. This allows vehicles to pass through with complete control of their inbound and outbound flow on a certain road leg. On the other hand, unsignalized intersections do not possess timed traffic lights. As mentioned in the study of Elefteriadou in 2014 [1], a stationary and untimed "stop" or "yield" signage controls this intersection and, thus, experiences more issues than a signalized junction.

One of the most common issues experienced in road intersections is vehicular congestion. To address this issue, several authors have been conducting studies that will manage traffic flow. Of the various intersections existing in worldwide roadways, this study will only focus on t-intersections. It comprises three road legs where a minor road

S.-H. Sheu (Ed.): ICIEA-EU 2024, LNBIP 507, pp. 71–82, 2024.
https://doi.org/10.1007/978-3-031-58113-7_7

meets a major road perpendicularly. Pananun et al. [2] define this as the type of junction that can handle low to medium traffic. The US Federal Highway team [3] characterizes t-intersections as fragments of remodeled crossroads.

Though researchers have set the assumption that t-intersections will help address and lessen existing conflict points at crossroads, this type of intersection has issues of its own. T-intersections are often constructed and operational without timed traffic signals, assuming motorists know the right-of-way. Bonela & Kadali [4] says that with t-intersections left unsignalized, traffic rules are less respected, eventually leading to congestion and conflicts. Addressing these problems, several authors, such as Surisetty & Sekhar [5], developed studies that can mitigate traffic flow. One significant solution is the integration of traffic lights. In 2017, [5] conducted a case-study experiment to generate more appropriate traffic signal timing at a t-intersection in India. The red, amber, and green times were optimized to lessen road congestion and problems at conflict points.

Conflict points are specific locations in an intersection where merging, diverging, or crossing problems occur [6] - where collisions and build-up are most likely to occur. Various researchers have studied conflict points in a t-intersection. APSEd India [7] published a guide on calculating conflict points within a t-intersection. Upon observation and plotting of traffic component behaviors traversing through a t-intersection, they have identified three (3) crossing conflicts, three (3) merging conflicts, and three (3) diverging conflicts. These points are applicable for t-intersections both with and without timed traffic lights. According to Alkaissi [8], this intersection type has six (6) existing movements or phases. Considering road configurations, vehicular volumes of the major and minor lanes are to be determined. The phase planning process will allow easier traffic management addressing the aforementioned conflict points.

This study will contribute to the academic community, especially to fields of engineering tackling road traffic management. This paper will fill in the limitations of existing literature that will address and eventually contribute to completing the most effective solution for road congestion, such as failing to integrate road configuration, phase planning, and conflict points at once. As for the industry, this study will address the high demand for land usage in global trading. Congestion rates will heighten, and solutions for managing traffic are needed. The transportation industry may be given opportunities to execute more traffic management projects. Disruptions and delays in trade and transport will be lessened, allowing industries to function well and meet demands.

2 Literature Review

2.1 T-Intersections

Pananun et al. [2] have characterized t-intersections as the type of road junction allowing a minor road to meet a major one at a right angle. Only low to medium traffic can only be handled by this intersection type. Authors Bonela & Kadali (2022) [4] have observed that, in t-intersections, priority traffic rules (e.g., right of way) are usually less respected and nearly disregarded. This problem often causes conflict and congestion in t-intersections. However, with the manageability potential validated by the FHART team (2004) [9], ways of addressing these problems can be developed.

Falling into a case study type, Surisetty & Sekhar [5] have studied the isolated case of the Kothavalasa t-intersection in India. Problems such as accidents, conflicts, and congestion have continuously risen in this signalized t-intersection. The Kothavalasa entails the major road configuration, considering it one of India's Major District Roads (MDR). Connecting three areas of Vizianagaram, Visakhapatnam, and Araku, three (3) traffic lights are also currently installed. Using a three-phase signal timing design, the authors have used Webster's Method to calculate the optimal signal cycles involving minimum vehicle delays per road leg. Actual data were collected and analyzed through hourly variations. These fluctuations aided in the traffic light timing calculations and eventually helped the generation of values for each traffic light color.

2.2 Traffic Signal Timing Optimization

Developing an optimized traffic signal timing system has been one of the solutions generated by authors wanting to address vehicular congestion at road intersections. A group of researchers, Ma et al. [10] and Yu et al. [11], conducted two (2) studies in 2020 managing isolated intersections by optimizing green light timing and considering vehicular traffic without road configurations. The first article published in 2020 [10] met the objective of maximizing the green light time for eight phases in a crossroad type of intersection. As an output, the researchers developed two sub-cycles for the 8-phased crossroad traffic, having an equally distributed green split timing. This study did not consider the exact red light timing scheme while vehicles are on "stop." Later in the same year, Yu et al. [11] addressed the same congestion problem at a crossroad intersection by calculating the optimal green splits for the minimum and maximum cycle lengths. The researchers have recommended varied green splits for different day peaks (morning, off-peak, and evening peak). A similar study was conducted by Li & Liu (2021) [12], now calculating the splits of all the red, amber, and green traffic light timing. The traffic scheme considered in this study includes three phases. In 2022, Li et al. [13] also crafted a complete traffic light timing optimization scheme. However, these studies failed to address the presence and significance of conflict points in timing plans.

3 Mathematical Modeling

The mathematical model will feature a simple single objective optimization model that will minimize the cycle time of the t-intersection traffic light system. In this case, the base model will only be defined by singular lanes for the major and minor roads of the t-intersection. For maximum navigation throughout the system, the mathematical model will also include several constraints and parameters for the latter scenario and sensitivity analyses. This mathematical model will then be coded as a genetic algorithm program at MATLAB R2020b (Table 1).

Table 1. Model Variables

Indices	
x	Traffic light unit
y	Road lane
z	Subphase index
Decision Variables	
Gxy	Green light time on traffic light x at leg y
R_{xy}	Red light time on traffic light x at leg y
A_{xy}	Amber light time on traffic light x at leg y
D_{yz}	Delay at leg y taking subphase z
Parameters	
NV_{yz}	Normal (min.) vehicle capacity at leg y taking subphase z
SV_{yz}	Saturated (max.) vehicle capacity at leg y taking subphase z
L_{xy}	Lost time at leg y under subphase z
AV_{yz}	Actual road volume at leg y taking subphase z

This study will be under an objective function minimizing the total cycle time in a t-intersection traffic system. The cycle time will be the summation of the green, red, and amber timing for each t-intersection lane, all measured in seconds. One whole cycle will be from the first color duration of the first traffic light unit until the end of the last traffic light time.

$$Min\ C_O = \sum G_{xy} + R_{xy} + A_{xy} \tag{1}$$

The objective function will be subject to several constraints. The first constraint (2) works for the green light timing decision variable. Allocating the number of seconds for the green time will be the ratio of normal and saturated vehicle capacities at individual lanes multiplied by the lost time per lane. Lost time is the duration wherein no passing vehicles in a certain lane could be observed.

$$G_{xy} \leq \frac{NV_{yz}}{SV_{yz}} L_{yz} \tag{2}$$

As for the following constraints, these will now account for the red and amber light timing decision variable. Since the timing of the traffic light colors will be highly dependent on the overall cycle time, the red light timing will be less than or equal to the difference between the total green light timing within the system and the green light timing of the assigned traffic light unit. Then, the amber light timing should be less than or equal to the total lost time in all the road legs.

$$R_{xy} \leq \sum G_{xy} - G_{xy} \tag{3}$$

$$A_{xy} \leq \sum L_{xy} \tag{4}$$

The delay constraint (4) carries the ratio of the actual road volume and the maximum vehicle lane capacity at a certain road leg. This decision variable will result in a percentage value.

$$D_{yz} \geq \frac{AV_{yz}}{SV_{yz}} \times 100\% \tag{5}$$

These constraints will be taken into account in genetic algorithm programming in MATLAB. A fitness function will be used to determine the goodness of fit of the targeted solution in terms of aligning it with the formulated problem. All iterations at the end of running the code will fit the population's required distribution. Further, a roulette wheel code will mutate or loop the iteration procedure and select combinations of the best probable solutions. After 25 iterations, the best green light timing to be used under a 120-s cycle time is shown in the table below (Table 2).

Table 2. MATLAB Model Green Timing Results

Input Direction	Duration (in seconds)
East	36
West	40
South	35

4 Scenario Analysis

4.1 Base Scenario

Before incorporating the values from the model, the base model will be an unsignalized t-intersection, taking all the turning behaviors from the 6-phase plan.

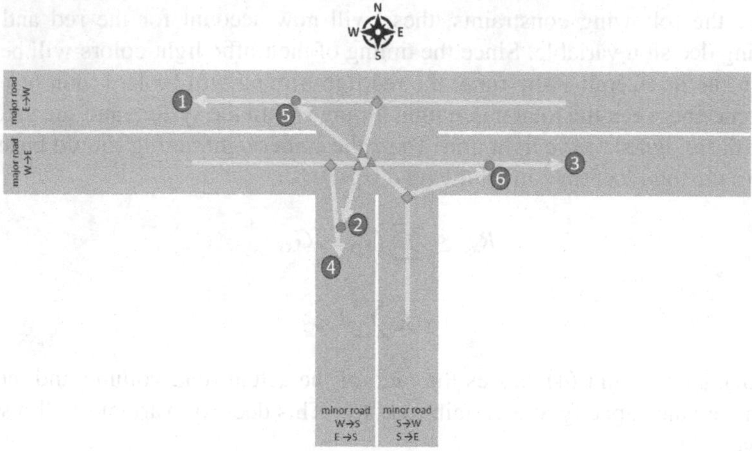

Fig. 1. Base Model Road Layout

Figure 1 shows the layout of the unsignalized base model. The phase plans are labeled with numbers. The conflict points are also plotted with orange triangles (crossing conflict), purple diamonds (diverging conflict), and green circles (merging conflict). With the initial simulation of the base system, density, delay, and volume values were obtained using PTV Vissim 2023. These are the available simulation result values that can be generated, which impose high relevance to the problem and goal of reducing congestion rates at t-intersections (Table 3).

Table 3. Base Model Characteristics and Results Summary

Component	E → W EMajor Road	W → E Major Road	W/E → S Minor Road	S → W/E Minor Road
Vehicle Input	1,500 veh./hr	1,500 veh./hr	×	400 veh./hr
Average Density	53.71	51.86	34.28	15.00
Average Delay	87.34%	42.74%	87.34%	23.03%
Average Volume	288	445	288	347

Progressing from the unsignalized base t-intersection model, the scenarios will feature three (3) signalized t-intersections. Traffic light timing values will be plugged into the systems built into the traffic simulator. The t-intersection system scenarios will demonstrate variations in phase planning, vehicle volume, and conflict points. This analysis will evaluate and compare each intersection lane's vehicle density, delay, and passing vehicle quantity (volume). The best scenario and policy will be the one that constitutes the least chance of having road congestion.

4.2 Scenario 1: 6-Phase Plan

The first scenario presents a t-intersection, with one (1) lane for each road leg, having all traffic light units activated. All lanes will have their respective vehicle inputs, taking in all behavioral turns and conflict points from the standardized 6-phase plan (Fig. 2).

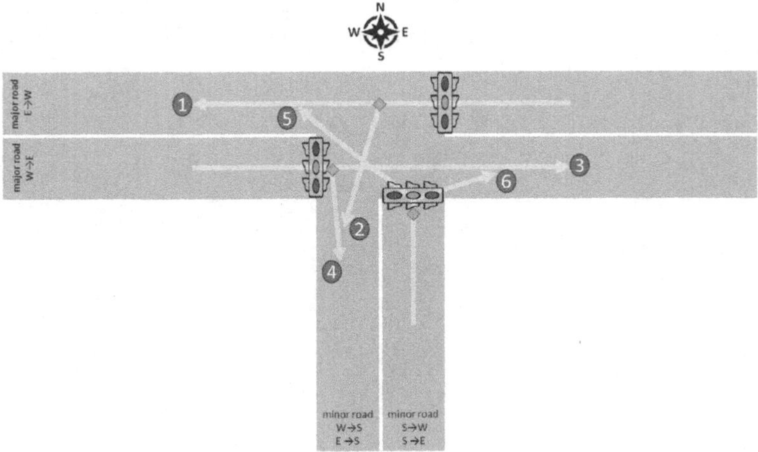

Fig. 2. Scenario 1 Road Layout and Phase Plan

Since all the traffic lights are activated, the complete traffic light phase plan will be applied. This scenario will also feature no overlapping green light timing (no inter-green timing). The lanes will commence vehicle movement, passing individually to avoid crossing conflicts. After running the simulation, density, delay, and volume results were gathered and presented in Table 4. As for this case, the lane with the most density and delay values is the south-to-west/east minor road. This lane has the highest chance of getting congested. Regarding the volume or service rate of passing vehicles, the west/east-to-south minor was the most efficient.

Table 4. Scenario 1 Characteristics and Results Summary

Component	E → W Major Road	W → E Major Road	W/E → S Minor Road	S → W/E Minor Road
Vehicle Input	≥1,500 veh./hr	≥1,500 veh./hr	×	≥400 veh./hr
Green Light Time	43 secs.	31 secs.	×	36 secs.
Cycle Time	120 secs.			
Average Density	61.75	63.74	13.71	84.72
Average Delay	48.18%	49.88%	0.17%	88.15%
Average Volume	440	322	474	327

4.3 Scenario 2: 5-Phase Plan

In this scenario, the left turning behavior from the east-to-west major lane is prohibited. This scenario seeks to smoothen the traffic flow and lessen the conflict point. The traffic system is illustrated below (Fig. 3):

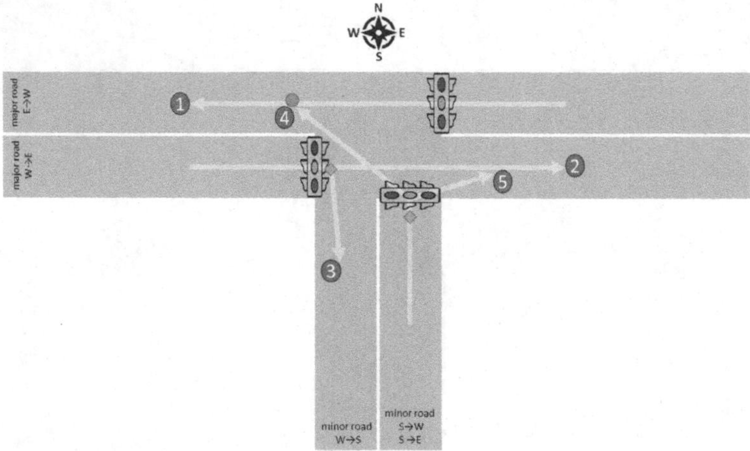

Fig. 3. Scenario 2 Road Layout and Phase Plan

The east-to-west traffic light unit can now go simultaneously with the other units. In addition, to maximize the simultaneous green light traffic, the total green light timing for each traffic light unit is split into halves for each lane. The least dense and delayed lane is still the west-to-south minor road, considering that the vehicle input source of this lane is decreased to just one lane. Then, the east-to-west major road has significantly increased its service rate for passing vehicles because the vehicles on this lane can pass freely without considering conflicting traffic (Table 5).

Table 5. Scenario 2 Characteristics and Results Summary

Component	E → W Major Road	W → E Major Road	W/E → S Minor Road	S → W/E Minor Road
Vehicle Input	≥1,500 veh./hr	≥1,500 veh./hr	×	≥400 veh./hr
Green Light Time	21, 22 secs.	15, 16 secs.	×	18, 18 secs.
Cycle Time	92 secs.			
Average Density	56.35	59.29	8.36	57.21
Average Delay	41.28%	48.39%	0.12%	81.53%
Average Volume	1,105	440	267	312

4.4 Scenario 3: 4-Phase Plan

The third scenario shows that two (2) left-turning behaviors are now restricted, reducing the phase plan to 4. In this case, more conflict points are lessened, expecting a better flow of traffic and a more efficient and strategic traffic signal timing plan (Fig. 4).

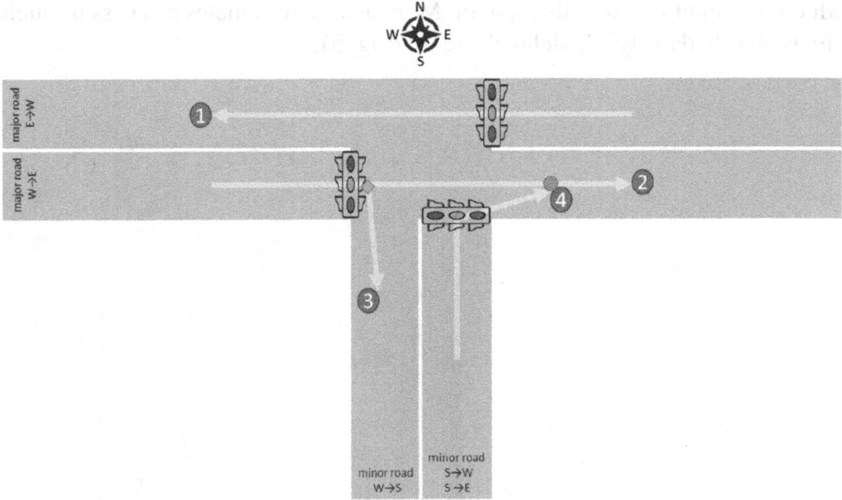

Fig. 4. Scenario 3 Road Layout and Phase Plan

Since two (2) turning behaviors are deduced from this scenario, all lanes can go simultaneously, increasing passing and service efficiency. The green timings are halved for equal distribution of the green timing from the optimization. In this case, the cycle time is also reduced to 92 s. At the end of the simulation, the west-to-south minor road still stands as the least congested road, having a low density, delay, and a high volume (service rate) for vehicles. The east-to-west major road also garnered a notably low delay and high vehicle passage, as its left-turning behavior is prohibited (Table 6).

Table 6. Scenario 3 Characteristics and Results Summary

Component	E → W Major Road	W → E Major Road	W/E → S Minor Road	S → W/E Minor Road
Vehicle Input	≥1,500 veh./hr	≥1,500 veh./hr	×	≥400 veh./hr
Green Light Time	21, 22 secs.	15, 16 secs.	×	18, 18 secs.
Cycle Time	90 secs.			
Average Density	29.62	53.42	15.45	49.01
Average Delay	4.99%	42.85%	0.12%	75.45%
Average Volume	1,459	767	492	316

4.5 Results Analysis

With the transition from an unsignalized t-intersection to a signalized system under three (3) different scenarios, a significant decrease has been observed in the density and delay values. The congestion rate becomes less probable after lessening conflict points in each scenario. In line with this, the volume or vehicle service rate has significantly increased by reducing conflict points in the system. More and more vehicles can pass through the system as vehicle density and delay decrease (Fig. 5).

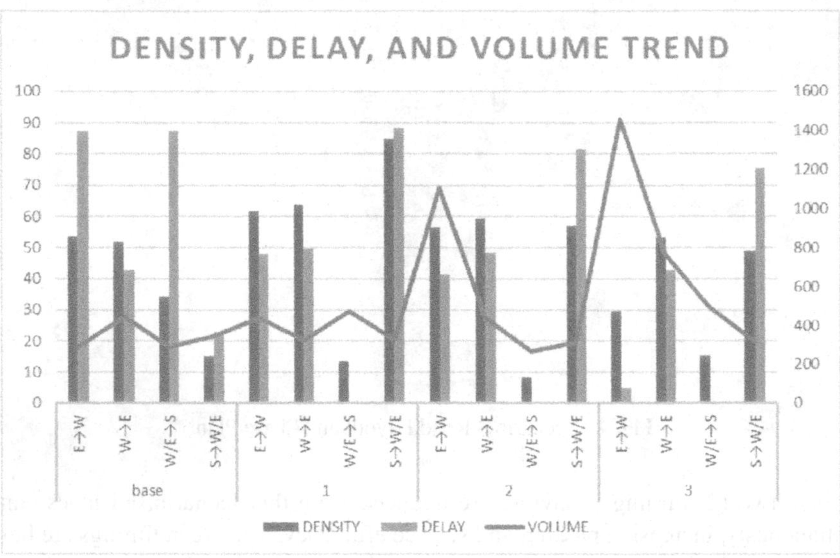

Fig. 5. Density, Delay, and Volume Results Trend Summary

The study also validated the significance of implementing phase planning adjustments to minimize the cycle time. Conflict points have also been reduced with the reduction of the vehicle turning phases. Since the green light timing of each of the scenarios with lessened conflict points can commence their traffic light units together, the cycle time was compressed (Fig. 6).

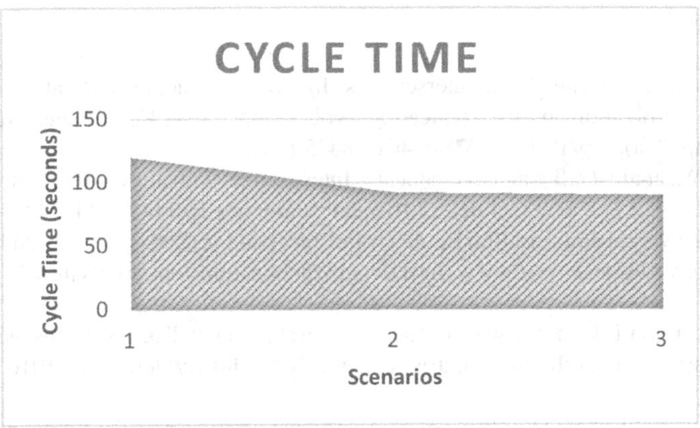

Fig. 6. Cycle Time Trend

5 Conclusions and Recommendations

There is a great chance for t-intersections to improve traffic flow by maximizing their potential to operate with traffic signals. As for traffic engineers, even under a low vehicle volume, they may opt to allow t-intersections to be signalized upon construction under both efficiency and safety considerations. The cycle time has significantly decreased by reducing conflict points, especially crossing conflicts. In conclusion, the study successfully introduced timing signal schemes that will work for both cases of a saturated and unsaturated t-intersection road with one (1) lane each for both the major and minor road legs. As per the results, the fourth scenario generated the slightest chance of congestion rate with the least cycle time. Under the consideration of prohibiting turning behaviors for vehicles in the targeted policy for implementation, the study could be further extended by introducing alternative turning points for the vehicles to still be able to access their desired destinations. Significant and valid observations were also recorded regarding a reducing trend in congestion rate (with average density and delay results) and an increasing trend in vehicle service rate per lane while minimizing cycle time.

This study then contributes more resources and avenues for studying a signalized t-intersection road to the academe. For future expansions of this traffic system management, the methodology can be extended in terms of its applicability and flexibility to other road intersection types and the increased number of traffic lights in the system. As per its contribution to the industry, the researcher recommends that traffic engineers implement a t-intersection system with lessened conflict points to maximize the service rate of passing vehicles for each lane. Since lessened conflict points prohibit a couple of turning behaviors for vehicles, traffic engineers and managers can implement future alternative turning behaviors away from the intersection point, still avoiding crossing conflicts in the middle of the system. In line with this, the recommended traffic signal timing for each traffic light unit can also be implemented with the minimized cycle time to lessen and avoid vehicle congestion, despite including pedestrian crossing. Upon construction, t-intersections are highly recommended to operate through a signalized traffic system.

References

1. Elefteriadou, L.: Unsignalized Intersections. In: An Introduction to Traffic Flow Theory. Springer Optimization and Its Applications, vol. 84, pp. 219–232. Springer, New York, NY (2014). https://doi.org/10.1007/978-1-4614-8435-6_10
2. Pananun, W., et al.: Traffic management at T Intersections with always-thru traffic. Int. Trans. J. Eng. Manag. Appl. Sci. Technol. (2018). https://doi.org/10.14456/ITJEMAST.2018.41
3. Federal Highway Administration Research and Technology. (2004). Signalized Intersections: Informational Guide. https://www.fhwa.dot.gov/publications/research/safety/04091/10.cfm#tab65
4. Bonela, S., Kadali, B.: Review of traffic safety evaluation at T-intersections using surrogate safety measures in developing countries context (2022). https://doi.org/10.1016/j.iatssr.2022.03.001
5. Surisetty, R., Sekhar, S.: Designing of a traffic signaling system at T-intersection. Int. J. Eng. Res. Appl. (2017). https://www.ijera.com/papers/Vol7_issue4/Part-3/M0704038286.pdf
6. Crowe, E.: Traffic Conflict Values for Three-Leg, Unsignalized Intersections. Transportation Research Record (2019). https://onlinepubs.trb.org/Onlinepubs/trr/1990/1287/1287-019.pdf
7. APSEd.India (2022). Types and Calculation of Conflict Points. Conflict Points at an Intersection
8. Alkaissi, Z.: Elements of Intersection Design and Layout. Traffic Engineering Lectures (2021). https://uomustansiriyah.edu.iq/media/lectures/5/5_2021_09_23!09_03_16_AM.pdf
9. Federal Highway Administration Research and Technology. (2004). Traffic Signal Timing Manual
10. Ma, W., et al.: Multi-objective optimization of traffic signals based on vehicle trajectory data at isolated intersections. Transp. Res. Part C: Emerg. Technol. (2020).https://doi.org/10.1016/j.trc.2020.102821
11. Yu, C., et al.: A time-slot based signal scheme model for fixed-time control at isolated intersections. Transp. Res. Part B: Methodol. (2020). https://doi.org/10.1016/j.trb.2020.08.004
12. Li, Z., Liu, H.: Signal timing optimization of shifted left-turn intersections. In: IOP Conference Series: Earth and Environmental Science (2021). https://doi.org/10.1088/1755-1315/791/1/012115
13. Li, W., et al.: Human-vehicle intersection traffic lights timing optimization research. J. Adv. Transp. (2022). https://doi.org/10.1155/2022/5549454

Lateral Stability Control for Four Independent Wheel Vehicles Considering the Surrounding Condition

Jinghua Zhang, Lifeng Ding, Junjian Chen, and Lei Yue(✉)

Guangzhou University, Guangzhou, China
{zjhjd,dinglf,2007700060,leileiyok}@gzhu.edu.cn

Abstract. During the steering manoeuvre, due to the variations such as vehicle parameters, speed and tire grip, vehicles may undergo understeer or oversteer, which may lead to vehicle deviation from expected path and lose of control, greatly threatening passengers' safety. Therefore, this paper proposes a coordinated control method for the lateral stability of four wheel independent steering vehicles with the surrounding condition considered, which could be adaptable to the surrounding environment and has excellent control performance. Firstly, a four independent wheel vehicle model combining longitudinal and lateral force is established. Secondly, a LQR controller with its parameters adjustable according to the surrounding condition is designed. Finally, the effectiveness of the control algorithm is verified via simulation.

Keywords: Four independent wheels · Lateral stability control · State feedback · LQR

1 Introduction

In the driving assistance system, the yaw stability control system is of great significance. It defines three operations during the steering process based on the vehicle's steering angle and lateral acceleration: oversteer, neutral steering, and understeer [1]. When the steering angle is excessive and on roads with low friction coefficient, the lateral force generated is relatively small, resulting in understeer. However, when driving at high speeds, excessive turning angles will generate excessive lateral force, leading to oversteer.

The yaw moment can be generated through a variety of ways, including tire lateral force generation (active front wheel steering, active rear wheel steering and all-wheel steering), tire longitudinal force generation (differential braking, differential drive), and combinations of any of the above. Furthermore, various

This research was funded by the Guangdong Provincial Junior Innovative Talents Project for Ordinary Universities (2023KQNCX054), Guangzhou Science and Technology Plan Project (202201010475, 202201020173) and National Innovation Training Project of Guangzhou University (No. 202211078101).

S.-H. Sheu (Ed.): ICIEA-EU 2024, LNBIP 507, pp. 83–94, 2024.
https://doi.org/10.1007/978-3-031-58113-7_8

tracking control methods such as inversion [2], adaptive tracking [3], neural networks [4], adaptive robust control [5], and other nonlinear analysis methods [6] can be easily applied to the control of yaw rate.

A substantial amount of research has been undertaken on yaw stability in recent years. Primarily focusing on lateral and yaw motions, existing methods like direct yaw-moment control (DYC) or active front steering (AFS) are no longer effective for preventing understeer. Consequently, some new research has emerged. [7] highlighted that achieving tighter turns necessitates a reduction in speed. They proposed an electronic stability control (ESC) integrated with an enhanced under-steering control (EUC) strategy, utilizing EUC for deceleration. [8] devised an optimal torque vectoring strategy tailored for four-wheel-independent-drive (4WD) electric vehicles to enhance lateral stability. On the other hand, [9] reframed the issue of understeer prevention by focusing on minimizing trajectory errors. These studies employed separate methodologies such as the P-control method and optimal control in their approaches. However, these works mentioned presume a separation between longitudinal and lateral forces. In practice, these forces are interconnected and governed by the constraints of the friction circle.

Therefore, this article proposes a dynamic model that integrates lateral and longitudinal forces, in order to avoid the problem of exceeding the friction circle limit caused by decoupling control of lateral and longitudinal forces. In addtion, In addition, it is proposed to link the coefficient matrix R of LQR control with the surrounding environment, and flexibly select control strategies based on the environment to achieve yaw stability control. The paper is organized as follows. Section 2 presents the combined longitudinal and lateral force dynamic model. In Sect. 3, an LQR controller and an R matrix tuning strategy is designed. Finally, simulation results are provided.

2 Four Independent Wheel Vehicles Model

To characterize the lateral motions of the vehicle, based on Newton's second law, the vehicle dynamics model is expressed as follows (Fig. 1):

$$\begin{cases} m\left(\dot{v_x}-\gamma v_y\right)=F_X, \\ m\left(\dot{v_y}+\gamma v_x\right)=F_Y, \\ I_z\dot{\gamma}=M_z, \end{cases} \tag{1}$$

where, F_X, F_Y, M_z represents the longitudinal force, lateral force, and rotational moment. When $\delta_f, \delta_r < 5°$, it can be approximated as $\cos\delta_f \approx \cos\delta_r \approx 1$, $\sin\delta_f \approx \sin\delta_r \approx 0$. It can be calculated that:

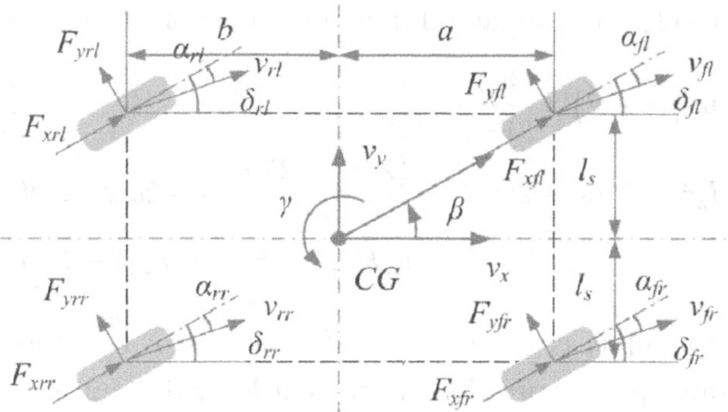

Fig. 1. Schematic diagram of four-wheel steering model

$$\begin{cases} F_X = F_{x_{fl}} + F_{x_{fr}} + F_{x_{rl}} + F_{x_{rr}}, \\ F_Y = F_{y_{fl}} + F_{y_{fr}} + F_{y_{rl}} + F_{y_{rr}}, \\ M_Z = \left(F_{xfr} + F_{yfr}\right) l_s - \left(F_{xfl} + F_{xrl}\right) l_s \\ \quad + \left(F_{yfl} + F_{yfr}\right) a - \left(F_{yrl} + F_{yrr}\right) b, \end{cases} \tag{2}$$

where, m is the total mass of the vehicle and I_z is the moment of inertia of yaw; a and b represent the distance from the front and rear axles to the center of gravity (CG), and ls represents half wheelbase; γ represents the yaw rate of the vehicle; v_x and v_y is the longitudinal and lateral speed velocity, respectively; $F_{x_{ij}}$ and $F_{y_{ij}}$ $(i, j = fl, fr, rl, rr)$ represent the longitudinal and lateral forces acting on the four wheels, respectively.

For the convenience of controller design, it is usually assumed that the stiffness of each wheel is constant, and the lateral force is represented as:

$$F_{yij} = 2c_{ij}\alpha_{ij}, \tag{3}$$

where, c_{ij} represents the lateral stiffness of the four wheels, α_{ij} represents the side slip angle of the four wheels. Based on vehicle kinematics, α_{ij} can be established by the following equation:

$$\begin{cases} \alpha_{fl} = \beta + \dfrac{a\gamma}{v_x} - \delta_{fl}, \\ \alpha_{fr} = \beta + \dfrac{a\gamma}{v_x} - \delta_{fr}, \\ \alpha_{rl} = \beta - \dfrac{b\gamma}{v_x} - \delta_{rl}, \\ \alpha_{rr} = \beta - \dfrac{b\gamma}{v_x} - \delta_{rr}, \end{cases} \tag{4}$$

where, δ_{ij} is the rotation angle of the four wheels; β is the lateral deviation angle of the center of mass, which is calculated by the following equation:

$$\beta \approx \tan\beta = \frac{v_y}{v_x}. \tag{5}$$

According to (1)–(5), the lateral dynamics of a vehicle can be expressed as:

$$\begin{cases} m\left(\dot{\beta}+\gamma\right)=\dfrac{4c\beta}{v_x}+\dfrac{1}{{v_x}^2}\left(2ca-2cb\right)\gamma-\dfrac{2c\delta_f}{v_x}-\dfrac{2c\delta_r}{v_x}, \\ I_z\dot{\gamma}=\left(2ac-2bc\right)\beta+\dfrac{\left(2a^2c+2b^2c\right)}{v_x}\gamma-2ac\delta_f+2bc\delta_r \\ \qquad +(F_{xfr}+F_{xrr}-F_{xfl}-F_{xrl})l_s. \end{cases} \tag{6}$$

Select control input $u=\left[\delta_f\ \delta_r\ F_{xfl}\ F_{xfr}\ F_{xrl}\ F_{xrr}\right]^T$, state variable $x=\left[\beta\ \gamma\right]^T$. The state space expression of the vehicle model can be obtained as:

$$\begin{bmatrix}\dot{\beta}\\ \dot{\gamma}\end{bmatrix}=A\begin{bmatrix}\beta\\ \gamma\end{bmatrix}+Bu, \tag{7}$$

where, $A=\begin{bmatrix}\frac{2(c_f+c_r)}{mv} & \frac{2(ac_f-bc_r)}{mv^2}-1\\ \frac{2(ac_f-bc_r)}{I_z} & \frac{2(a^2c_f+b^2c_r)}{I_zv}\end{bmatrix}$, $B=\begin{bmatrix}-\frac{2c_f}{mv} & -\frac{2c_r}{mv} & 0 & 0 & 0 & 0\\ -\frac{2ac_f}{I_z} & \frac{2bc_r}{I_z} & -l_s & l_s & -l_s & l_s\end{bmatrix}$.

3 Lateral Stability Coordination Controller Considering the Surrounding Condition

3.1 Trajectory Tracking Controller

The performance of vehicle tracking expected trajectory can be evaluated by a objective function:

$$J=\frac{1}{2}\int_{t_0}^{t_f}\left[e^{\mathrm{T}}(t)Qe(t)+u^{\mathrm{T}}(t)Ru(t)\right]\mathrm{d}t, \tag{8}$$

where, t_0 and t_f are the start time and the end time respectively; Q is a $2*2$ semi definite state weighted diagonal matrix; R is a $6*6$ positive definite control weighted diagonal matrix; $e(t)=y_r-y$, is the tracking error.

In this paper $Q=diag(q_{11},q_{22})$, $R=diag(r_{11},r_{22},...,r_{66})$. The elements q_{11}, q_{22} represent the relative importance of each state error, and the elements $r_{11},r_{22},...,r_{66}$ represent the relative restraints of each input. The diagonal elements of R are adjusted according to the surrounding condition (specific adjusting strategies are introduced in Sect. 3.2).

According to Pontryagin's minimum principle, it can be concluded that the optimal control $u^*(t)$ is:

$$u^*(t)=-R^{-1}B^{\mathrm{T}}Px(t)+R^{-1}B^{\mathrm{T}}g, \tag{9}$$

where, P, g satisfies the following equation:

$$A^TP+PA-PBR^{-1}B^TP+C^TQC=0, \tag{10}$$

$$g\approx[PBR^{-1}B^T-A^T]^{-1}C^TQy_r. \tag{11}$$

Based on the above formula, the optimal trajectory can be obtained:

$$\dot{x}^*(t) = [A - BR^{-1}B^{\mathrm{T}}P]x + BR^{-1}B^{\mathrm{T}}g \tag{12}$$

3.2 Weighted Matrix Adjusting Strategy

When a car undergoes understeer and oversteer during the steering manoeuvre, the solutions include braking the outer wheels, driving the inner wheels, turning the front wheels outward, and turning the rear wheels inward. However, different control methods can have different impacts on the vehicle's trajectory (as shown in Fig. 2). Therefore, this article adjusts control parameters accroding to the surrounding condition. Based on the following three principles, the adjusting strategies are given in Table 1:

a. When the vehicle is in understeering/oversteering, the braking effect of the inner/outer wheels is the best [10].
b. Braking wheels is more stable than driving wheels because braking wheels can appropriately reduce vehicle speed.
c. Turning the front wheels is more stable than turning the rear wheels, and in addition, cars usually use front wheel steering.

After determining the adjustment strategy for the R matrix, the vehicle will adapt the diagonal elements of the R matrix based on its surrounding environment, thereby achieving flexible control. The control flow chart is illustrated in Fig. 3.

Table 1. Control strategy considering the surrounding vehicle condition

Right turn with oversteer or left turn with understeer $((\delta_{hw} \cdot (\lvert\gamma\rvert - \lvert\gamma_r\rvert) < 0))$		
surrounding condition	control method	R matrix assignment method
Clear behind	Brake the left wheel	$r_{11} = r_{22} = r_{44} = r_{66} = Lr_{33} = r_{55} = S$
Obstacle behind clear ahead	Drive the right wheel	$r_{11} = r_{22} = r_{33} = r_{55} = Lr_{44} = r_{66} = S$
Obstacle ahead and behind clear left	Front wheel left turn	$r_{22} = r_{33} = r_{44} = r_{55} = r_{66} = Lr_{11} = S$
Obstacle ahead behind and left clear right	Rear wheel turning right	$r_{11} = r_{33} = r_{44} = r_{55} = r_{66} = Lr_{22} = S$

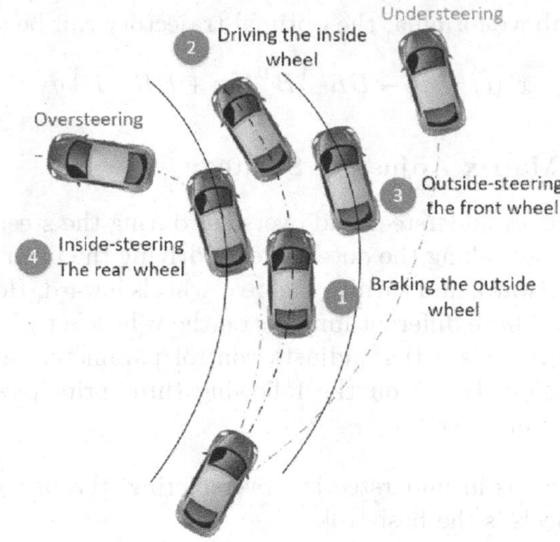

Fig. 2. Driving trajectories under different strategies

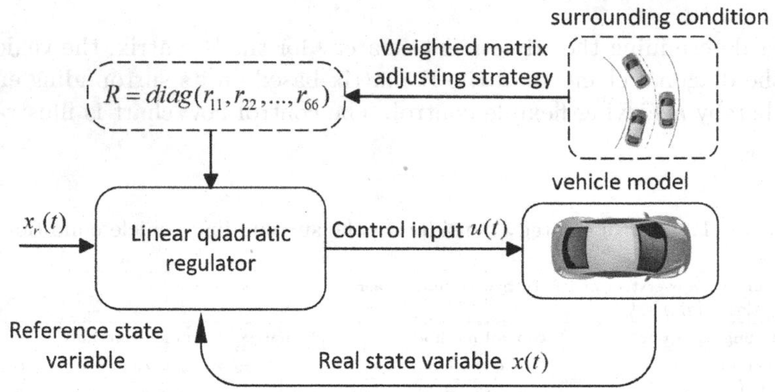

Fig. 3. Driving trajectories under different strategies

4 Simulation Verification

To verify the feasibility and effectiveness of the proposed algorithm, this paper employed both numerical simulations and MATLAB-CarSim coalition simulations are performed. In this work, objects within a distance of less than 5 m from the vehicle were considered obstacles, while objects beyond 5 m were considered clear.

4.1 Numerical Simulation

In this section, the reference model $\gamma_r = K_{sw}\frac{v}{L(1+K_f v^2)}\delta_{hw}$, $\beta = 0$ is used as the reference target, where, $L = a + b$, $K_f = \frac{m}{2(a+b)^2}(\frac{b}{c} - \frac{a}{c})$. The simulation parameters of the vehicle are shown in Table 2.

Table 2. Parameters of the vehicle

parameter	m	c	I_z	K_f	q_{11}, q_{22}
Value	1430 kg	-44020N/rad	2059.2kg/m^2	-0.0012	100
parameter	K_{sw}	L_w	v_x	S	L
Value	0.05	0.74 m	20 m/s	0.001	1000

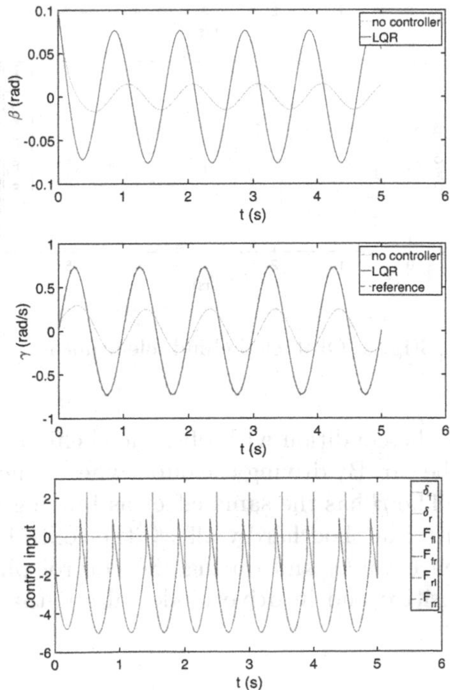

Fig. 4. Clear behind

In Fig. 4, a simulation of clear behind the car is presented. The sideslip angle is greater under LQR control than that without control. Because the vehicle undergoes understeering, controller interferes to correct the steering angle, thereby increasing the sideslip angle. In addition, it can be seen from the control input that four wheel control inputs are negative and in braking mode (alternating braking of left and right wheels). This demonstrates the effectiveness of the algorithm.

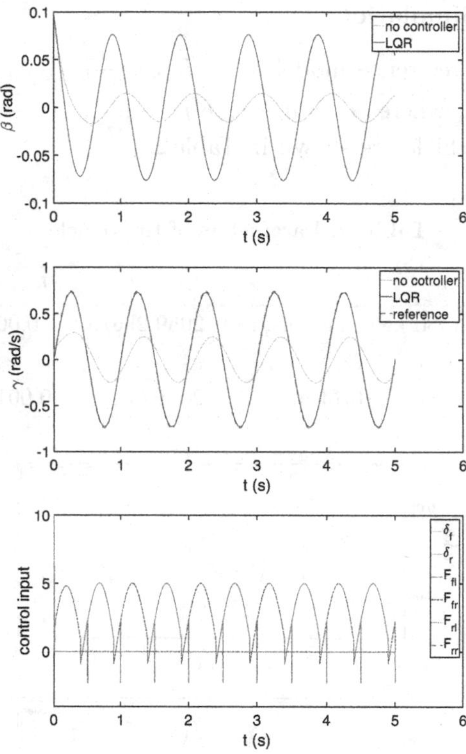

Fig. 5. Obstacle behind, clear ahead

Figure 5 simulates the condition with obstacles behind the car, and without obstacles in front of the car. By driving the outer wheels, the steering stability of the car is controlled. The β has the same effect as braking the external wheels, and the choice of control method hardly affects the control effect. The γ is well tracked to the reference value, and during the control phase, the four wheel control inputs are positive and in drive mode (alternating driving of left and right wheels).

4.2 MATLAB-CarSim Cosimulation

CarSim ensures high fidelity in vehicle dynamics and provides a rich testing environment. Therefore, a collaborative simulation between CarSim and MATLAB is conducted to validate the practicality of the proposed control strategy. In this simulation, we selected the C-Class Hatchback model with an initial velocity of 10 m/s. The road profile consists of a straight segment and a curved arc, with a road friction coefficient of 0.85. There are three simulation scenarios: no vehicle behind, no vehicle in front but a vehicle behind, and vehicles both in front and behind with no vehicle on the left.

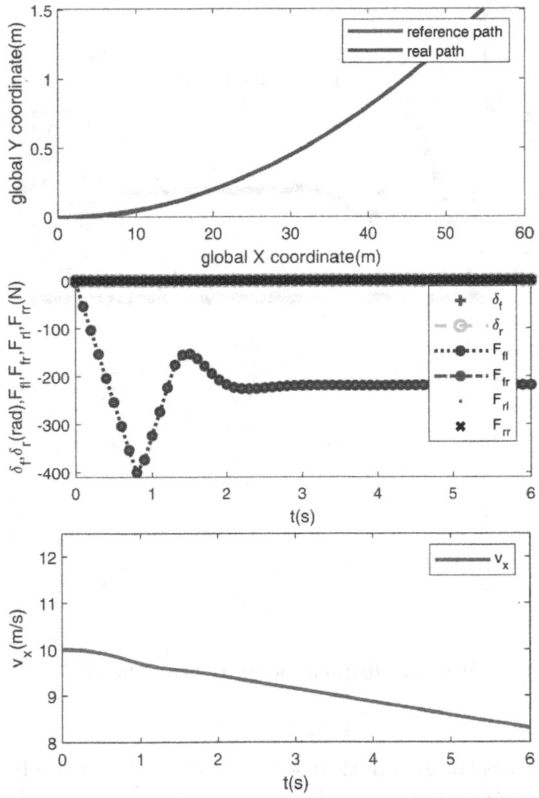

Fig. 6. Clear behind

Figure 6 depicts a simulation scenario in which the vehicle undergoes insufficient left-turn steering with no obstacles behind. In this case, steering control is achieved by braking the left-side wheels. It can be observed that the four control inputs, $\delta_f, \delta_r, F_{fr}, F_{rr}$ are nearly zero, since their corresponding elements in the R matrix are assigned larger values and thus heavily constrained. On the other hand, the two control inputs, F_{fr}, F_{rl} have significantly larger values, as their corresponding elements in the R matrix are assigned smaller values, encouraging their increase. Additionally, the reduction in vehicle speed caused by wheel braking also serves as another validation of the applicability of this control strategy in scenarios where there are no obstacles behind the vehicle.

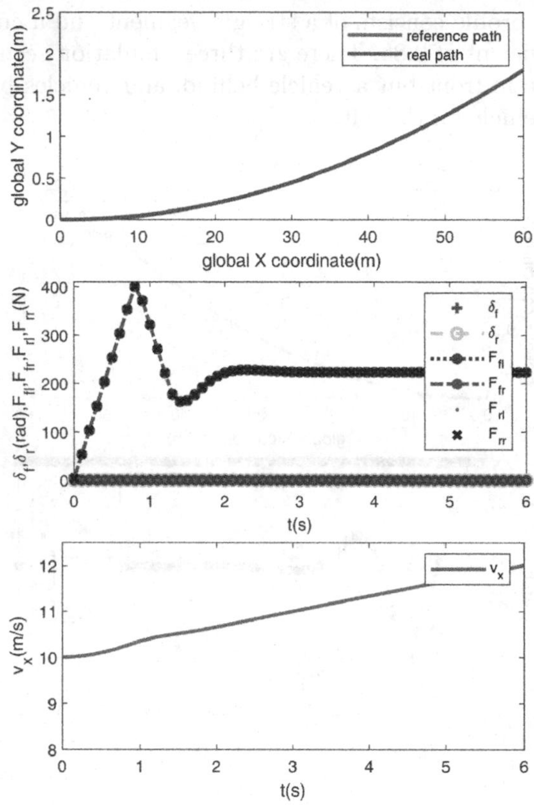

Fig. 7. Obstacle behind, clear ahead

Figure 7 illustrates the simulation scenario where the vehicle undergoes insufficient left-turn steering with obstacles behind and clear ahead. Due to the presence of obstacles behind, steering control is achieved by driving the right-side wheels. Similarly, the increase in vehicle speed is attributed to driving the right-side wheels.

Figure 8 depicts a simulation scenario where the vehicle undergoes insufficient left-turn steering with obstacles both ahead and behind, but with no obstacles on the left side. In this case, the steering control strategy takes turning the front wheels to the left, with the diagonal elements of the R matrix taking values of $r_{22} = r_{33} = r_{44} = r_{55} = r_{66} = L$. $r_{11} = S$. From the control inputs in the figure, it can be observed that only the control input corresponding to r_{11} is encouraged, while the other control inputs are suppressed. Through the simulation analysis of these three scenarios, the proposed method could select different control strategies based on environmental information to achieve steering control.

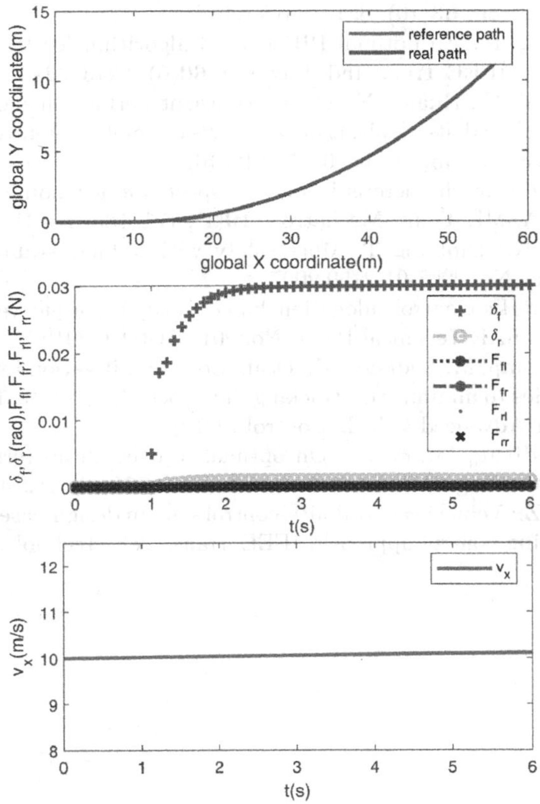

Fig. 8. Obstacle ahead and behind but clear left

5 Conclusion

This paper proposes a lateral coordination stability algorithm that considers the surrounding condition. By adjusting the elements in the matrix R of LQR controller according to the surrounding condition, the flexibility of the control strategy is achieved. Finally, the effectiveness and feasibility of the proposed algorithm are tested by the simulation. In future, we will consider how to achieve accurate control under the uncertainty of vehicle parameters and interference from external factors.

References

1. Frendi, S., et al.: Tracking controller design of a sideslip angle and yaw rate for electrical vehicle bicycle model. IFAC-PapersOnLine **49**(5), 169–174 (2016)
2. Sun, W., Zhang, Y., Huang, Y., et al.: Transient-performance-guaranteed robust adaptive control and its. Application to precision motion control systems IEEE Trans. Ind. Electron. **63**(10), 6510–6518 (2016)
3. Jing, X., Cheng, L.: An optimal PID control algorithm for training feedforward neural networks. IEEE Trans. Ind. Electron. **60**(6), 2273–2283 (2013)
4. Sun, W., Zhang, Y., Huang, Y., et al.: Transient-performance-guaranteed robust adaptive control and its application to precision motion control systems. IEEE Trans. Ind. Electron. **63**(10), 6510–6518 (2016)
5. Jing, X.: Nonlinear characteristic output spectrum for nonlinear analysis and design. IEEE/ASME Trans. Mechatron. **19**(1), 171–183 (2014)
6. Liebemann, E.K.: Dipl Ing T. More safety with vehicle stability control. SAE Technical Paper No. 2007-01-2759.2007
7. Kang, J., Heo, H.: Control allocation based optimal torque vectoring for 4WD electric vehicle. SAE Technical Paper No. 2012-01-0246.2012
8. Gordon, T., Klomp, M., Lidberg, M., et al.: Control mitigation for over-speeding in curves: strategies to minimize off-tracking. In: Proceedings of the 11th International Symposium on Advanced Vehicle Control (2012)
9. Klomp, M., Lidberg, M., et al. : On optimal recovery from terminal understeer. Proc. Inst. Mech. Eng. Part D: J. Automob. Eng. **228**(4), 412–425 (2014)
10. Zhou, H., Liu, Z.: Vehicle yaw stability-control system design based on sliding mode and backstepping control approach. IEEE Trans. Veh. Technol. **59**(7), 3674–3678 (2010)

Design of an Interactive Scheduling Heuristic-Based Application

Edmond Duay, Gene Mark Gondraneos, Karisha Ann Indino-Pineda, and Rosemary Seva(✉)

Industrial and Systems Engineering, De La Salle University, Manila, Philippines
rosemary.seva@dlsu.edu.ph

Abstract. Scheduling classes is an essential and challenging task due to academic rules and constraints imposed by the system. Previous studies have tackled the problem of class scheduling, but teacher schedule and course preference have not been incorporated. The usability of previous software developed were also not established. This study aims to create an automated tool that will minimize the time and errors the department heads encounter in creating a faculty schedule. The Interactive Scheduling Heuristic-based Application (ISHA) was developed using a user-centered design approach. Seventeen department administrators were interviewed to determine the tasks involved in the scheduling process, the constraints, and best practices. The contents of the software were determined from the outcome of a task analysis. The system was designed to be embedded in the web application and the algorithm itself when autogenerating a schedule. The algorithm prevents schedules from overlapping with one another. The usability of the application was assessed using efficiency by comparing the speed of scheduling with and without the tool. ISHA was able to ease the burden of the department heads in preparing a schedule that considers all the constraints imposed by the system. The scheduling task had been shortened, less prone to errors, and had been more oriented with the capabilities of the department heads.

Keywords: scheduling · system usability scale · heuristics

1 Introduction

Scheduling classes is an essential and challenging task in managing educational institutions. The difficulty lies in observing the academic rules and constraints of the system [1]. Recent solutions to the class scheduling problem used mathematical solutions such as genetic algorithm [2], linear programming [1], multi-agent system-based method [3], tabu search algorithm [4], combinatorial optimization [5], and divide and conquer algorithm [6], among others. Although the problems in class scheduling are similar, the objectives, resources, and constraints differ in context. The most common goal is to minimize conflicts [1], facilitate the process [3], reduce waiting time [7] and maximize the use of resources [1]. Mathematical tools enable optimization of resources given the system's objective. However, not everyone involved in the scheduling process is

S.-H. Sheu (Ed.): ICIEA-EU 2024, LNBIP 507, pp. 95–106, 2024.
https://doi.org/10.1007/978-3-031-58113-7_9

well-versed in mathematics. Departments in universities are diverse, and solutions need to consider the limitations of the users in terms of efficiency, memory, and flexibility. Thus, it is important to consider the context of the institution and make the solution more acceptable to the users.

Studies done on university resource scheduling considered only a few variables that are quantifiable. Can et al. [8], for instance, considered the preferences of instructors and avoided the overlapping of classes. A more comprehensive solution that considered the context of the school is the HORARIS system developed for the University of Valencia in Spain. The user can input the set the objectives and parameters and then the best solution is obtained using a heuristic [9]. HORARIS, however, did not consider the preference of the faculty in terms of scheduling which is a realistic constraint that should have been included. Commercial academic scheduling software available only include the scheduling of school resources, not necessarily classes.

Faculty scheduling, especially in a large department with more than 30 academic staff, is a tedious task usually assigned to a department administrator (DA). For De La Salle University, the scheduling process starts at least two months before the trimester and includes giving courses to academic staff and allocating rooms/laboratories. The schedule is summarized using a plantilla form containing the staff name, courses assigned, load in units, section, number of students, and room.

The constraints of maximum faculty load, number of preparations, faculty specialization, schedule preferences, and fatigue constraints complicate this task. A faculty can only handle a maximum of 15 units per term with a maximum of 3 preparations. They cannot teach for more than 4.5 consecutive hours to avoid fatigue and have specific preferences regarding time. Seventeen department administrators were interviewed, and some of the most common problems cited were scheduling mistakes such as assigning two courses in one classroom, assigning one faculty in one time slot, or violating established rules in loading. The schedule had to be revised several times because of a mistake in identifying the number of sections. Each change entails adjustment of faculty load and was done manually using pencil and paper. Experienced department heads still commit some errors, but the task is especially difficult for new department heads.

Previous studies have tackled the problem of class scheduling, but teacher schedule and course preference have not been incorporated. The usability of previous software developed were also not established. This study aims to create an automated tool that will minimize the time and errors the department heads encounter in creating a faculty schedule. The device will encapsulate the best practices in scheduling adopted by the department heads and allow them to make quickly as the need arises.

2 Method

The Interactive Scheduling Heuristic-based Application (ISHA) development followed a user-centered design approach shown in Fig. 1.

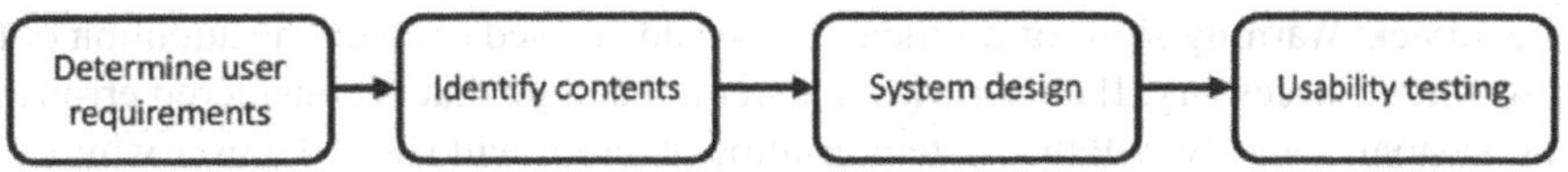

Fig. 1. ISHA design processs

2.1 Determine User Requirements

Seventeen department administrators (DAs) were interviewed to determine the tasks involved in the scheduling process, the constraints, and best practices. A task analysis was done to determine the specific scheduling steps. Some DAs were observed performing the tasks to ensure proper documentation. Constraints were obtained from document analysis, such as the policies and guidelines of the school and accrediting institutions. These constraints were also discussed by the DAs when they were queried about the process of allocating courses to faculty members. Best practices were identified by asking them for specific solutions to common problems in scheduling.

2.2 Identify Contents

The contents of the software were determined from the outcome of the task analysis. Specific contents were identified at each step from the beginning to the end of the process, incorporating the best practices obtained from the interviews. Faculty data such as units to be taught, specialization, and time preferences were obtained from the DA, while course data were obtained from the College Administrative Assistant (CAA), which keeps track of all the courses to be offered by the college. The number of sections was computed by the DA based on the number of students expected to enroll in each course.

2.3 System Design

The system was designed to be embedded in the web application and the algorithm itself when autogenerating a schedule. The algorithm prevents schedules from overlapping with one another. Errors are not allowed as the algorithm performs a series of checks to make sure that two different courses cannot be scheduled at the same time, on the same date, in the same room. The algorithm also considers policies while creating the schedule, so the burden of memorizing the rules is minimized. The web application was also designed to facilitate changes in schedule using the autogenerate function. A sudden change in information, like the removal of a professor because of availability issues, can be easily solved by the system. It will just create a new schedule without the professor once it is taken out by the user on the professor page. Finally, one other group of problems addressed by the design of the web application is the task conditions that the user experiences.

The design also considered human factors issues such as:

Memory: The knowledge-based system reduces the number of items the user must consider when scheduling.

Performance: Minimizes the time in which the process is completed.

Feedback: Warning signs or notifications would be used to catch the attention of the user whenever necessary. If a user tries to enter a schedule that is considered erroneous by the system, not only will the system disallow it, but it will show the user why.

Flexibility: The user can decide to bypass certain rules if needed.

Value-added features: An auto-save prompt and an undo function would be very user-friendly features.

Easy to Learn: Tutorials that are integrated into the interface for first-time users would make sure that everyone has the option to learn the basics.

Adjustability: Being able to customize the system goes beyond the data you put in. Customizing it means that the user may be able to turn some settings or features on and off. Pop-up warnings and the color of notifications may be changed to suit the users' preferences.

Figure 2 summarizes the conceptual design of ISHA.

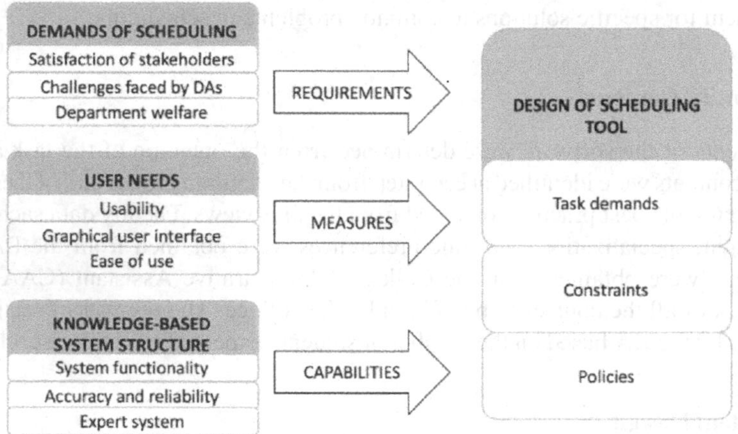

Fig. 2. Conceptual Design of ISHA

2.4 Usability Testing

Only one respondent was tested to evaluate the usability of the web application. The respondent involved is an incumbent DA. The test lasted for approximately one hour, including the briefing and the actual testing itself. The user was asked to perform specific tasks in accordance with the testing plan. The testing plan included using all the features of the web application, such as adding professors, courses, rooms, and time; browsing courses; changing the number of sections; assigning rooms to courses; assigning professors to courses; and generating a schedule.

The usability of the application was assessed using efficiency by comparing the speed of scheduling with and without the tool. The DA was also interviewed to gauge the overall usability of the tool.

3 Results and Discussion

3.1 User Requirements

The following issues were identified based on the interview with the DAs:

- The scheduling process is difficult to learn, especially for first-time users. It is very prone to error because of the constraints to be remembered by the DA
- The schedule should consider both the convenience of the faculty and students
- Any revision done will affect the whole schedule, so it is difficult to correct errors
- The process should be finished within a short time frame
- Many revisions need to be done until it is finalized because of uncertainties
- It is difficult to check errors
- The schedule is especially difficult to do for large departments because of the significant volume of resources to be managed

3.2 Software Algorithm

Two separate algorithms were developed: scheduling classes of the students from the department and from other departments. The algorithm assumed that the following information is available: an initial list of course offerings for the following term, available faculty members, number of teaching units available for each faculty member, faculty preference schedule, room, and laboratory availabilities.

The following are the best scheduling practices incorporated in the application's design:

- Computing for the total number of units of course offering and comparing it with the total units of faculty load available
- Considering hard constraints first in the scheduling process
- Pairing faculty with courses
- Scheduling by Student Batch
- Scheduling soft prerequisites back-to-back for irregular students
- Considering a backup plan in case courses are dissolved

Initialization steps of the algorithm. Only the user may be able to fill in the data in the three modules: Subject, Professor, and Room Lists. The web application has no capability to populate its own database for the three modules. The preference of the professors in the department is also something that the web application cannot predict or guess about. This is one of the inputs that only the department head can dictate or elaborate. Usually, the assigned department chair or vice-chair knew how professors were willing to be scheduled. Every term, there may be changes in the implemented school policies in terms of the scheduling of classes and even in the number of school days in a week. These are just some of the possible policies that may vary from one term to another, and with that, the settings of the web application must be changed. Some of the possible policies that may vary from one term to another have been incorporated in the design of web applications already, and those that are not would require the help of the developer.

Scheduling courses for students of the department. The DA prepares the schedule for each block of students on-track, which is done per batch. The DA can opt to pick a specific batch for scheduling first. After determining the batch to be scheduled, the system now checks for available timeslots that can be paired with the preference of the professor and the courses of that batch. If there is a course in that batch that matches the preference of the professor, then it is scheduled for the first timeslot, which is Monday morning. A timeslot is tagged as available if it satisfies the following conditions: there is no class scheduled, it is not blocked by any preference, and it does not violate a policy. The system schedules by first checking the timeslot, then the day, and then the room.

If there is a case that all combinations of timeslots, days, and rooms for a paired course-professor cannot be accommodated in a preferred timeslot, then the system would schedule the paired course-professor in a non-preferred timeslot. In a 6-day class-week, Monday is paired with Thursday. Therefore, for a regular twice-a-week three-unit course scheduled on a Monday, the paired course will reflect on Thursday as well. The schedule is done in the following order: department administrators, full-time faculty, and then part-time faculty. Higher-ranked faculty are given priority in scheduling. Lastly, for the given batch, repeater courses are now scheduled in the same order as how the regular courses were scheduled. The system then moves to the next batch and repeats this process. Table 1 summarizes the initialization process of the algorithm at the beginning of the term.

Scheduling courses for other departments. Service courses are those provided by one department to another. These courses are scheduled based on the request of the department. In this case, priority is given to professors who teach the course. The DA will first check if the schedule requested matches the availability of the professor. If there is a match, then the load is given to that professor. If there is no match, an alternative time is given to the professor that minimizes vacant time and does not violate any policy. The courses are then scheduled until all sections are done. If one professor is not enough to fulfill the requirement of the course requested in terms of units, then other professors will be given the load based on their time preference.

Schedule rating system. A schedule rating system was designed to evaluate the generated schedule considering all constraints. Two types of ratings were generated for each schedule: professor preference rating and student schedule rating. The student schedule rating is based on the desirability of the student schedule generated in terms of putting classes as close together as possible. Each professor was given a base score of 20, and every case violation merits a point deduction. The scoring system is based on professor feedback about what constitutes a good schedule. Professors prefer to have their teaching load clustered together during the day. Classes should also be scheduled consecutively with minimum vacant or breaks in between.

Three cases were considered for deductions, as shown in Table 2. The main difference between the scoring system of professors and students is that students do not have a set preference. Therefore, scoring is concentrated on how close their courses are to each other and the absence of large breaks. Only case 2 and 3 applies to students. Another difference is that students can have three consecutive courses scheduled, and deductions are only made for breaks more than two consecutive periods as opposed to professors that have score deductions for breaks more than one period.

Interface Design. Some of the screens designed for ISHA are shown in Fig. 3. The home page contains initialization tasks for the user to complete prior to scheduling. On this page, the user can select an academic term that enables the software to generate courses that need to be scheduled based on the student flowchart. It can also provide a list of projected courses for the term and a checkbox that shows if repeat courses are to be included in scheduling.

Table 1. Initialization Process

Step No.	Step Name	User Inputs	Objective/Constraints Defined
1	Select Term	Term to schedule	Determines the set of courses to be suggested in the following step; from the database of the courses/flowchart, the application automatically selects the set of courses regularly offered for the term and suggests major courses offered the term prior for the repeater section. The number of sections to be opened for a course must also be inputted in this section. This step also allows the user to re-use previously saved schedules by allowing the user to re-access the previously defined constraints, edit them, and automatically generate a schedule for the same term again
2	Select Courses	Courses offered	From the previous step, which automatically suggests courses regularly offered for the term selected, the user to verify the set of courses to be offered and activate courses to be offered to repeater sections
3	Select Rooms	Rooms available (Resource Constraint)	User defines the rooms available for use; if the entire room slots for the day or for the week are not exclusive to the department's use, the user may also define the time frame in which the rooms can be used. This will allow the application to assign rooms defined in this section to classes scheduled
4	Manage Sections	Sections opened	The section names for each year level are activated to assign to the courses offered; a course was defined to open two sections, and the system will get the name of the section through this constraint
5	Select Professors	Professor pool (Resource constraint)	Step 5 allows the user to select the faculty for the load, thus excluding those who are on leave and the like. In this section, the units available to teach for each professor must also be defined to avoid giving more load than the required teaching units. It also allows the user to define the courses that are preferred to be taught by the professor so the next step can define the courses the professor will teach for the term

(*continued*)

Table 1. (*continued*)

Step No.	Step Name	User Inputs	Objective/Constraints Defined
6	Manage Professors	Professors' preference (time and courses)	Two constraints concerning the faculty are in this step: the courses that a professor will teach during the term and the preferred time of the professor. The application generates a selection of courses that match the preference of the professor and are indicated to be offered for the term. This section allows the user to input the preferred time, optional time, and the teaching timeslot that is not preferred by the professor
7	Manual Scheduling	Pre-scheduled courses	This step lets the user create the schedule manually but with guidance so as to avoid policy violations and conflicts that may be committed. The application prevents the courses defined in this section from being part of the scheduled auto-generation

Repeat courses are those that are not on track during the term. The courses to be offered can still be changed, and the number of sections for each course can be assigned. The three buttons on the right of the home page are the load assignment button, the view faculty button, and the view list of rooms button. The load assignment button displays the page where the user can assign courses to the faculty, the room, and the section. The view list of faculty shows the current list of faculty where the user can add or delete faculty members, assign a load to different faculties, and choose which faculty members are to be included in the list for that term. The view list of rooms button permits the adding and deleting of rooms to be used in the scheduling.

Table 2. Deduction per Case

Deduction	Case 1: For every course not within the professor's zone of preference	Case 2: More than one break in between clusters	Case 3: Not clustered together when possible
−1	One timeslot outside the nearest point in the zone of preference	Two timeslots serve as a break	For every course not belonging to a cluster, whenever possible
−2	Two timeslots outside the nearest point in the zone of preference	Three timeslots serve as a break	
−n	N timeslots outside the nearest point in the zone of preference	N + 1 timeslots serve as a break	

The input provided on the home page enables the system to automatically specify the set of courses per batch and by status of the student. Courses are classified as regular (on-track), repeat, and service (offered to other departments). Each course requires manpower and is accounted for in the scheduling process. Faculty members are required to carry a minimum of 12 units load every term that includes teaching, research, and administration.

Once the number of units and the faculty teaching load units are compared and balanced by the system, the assignment of load follows. The immediate page right after the initialization page is the Faculty Loading List page. Each faculty must be assigned a number of courses to fulfill the minimum load requirement.

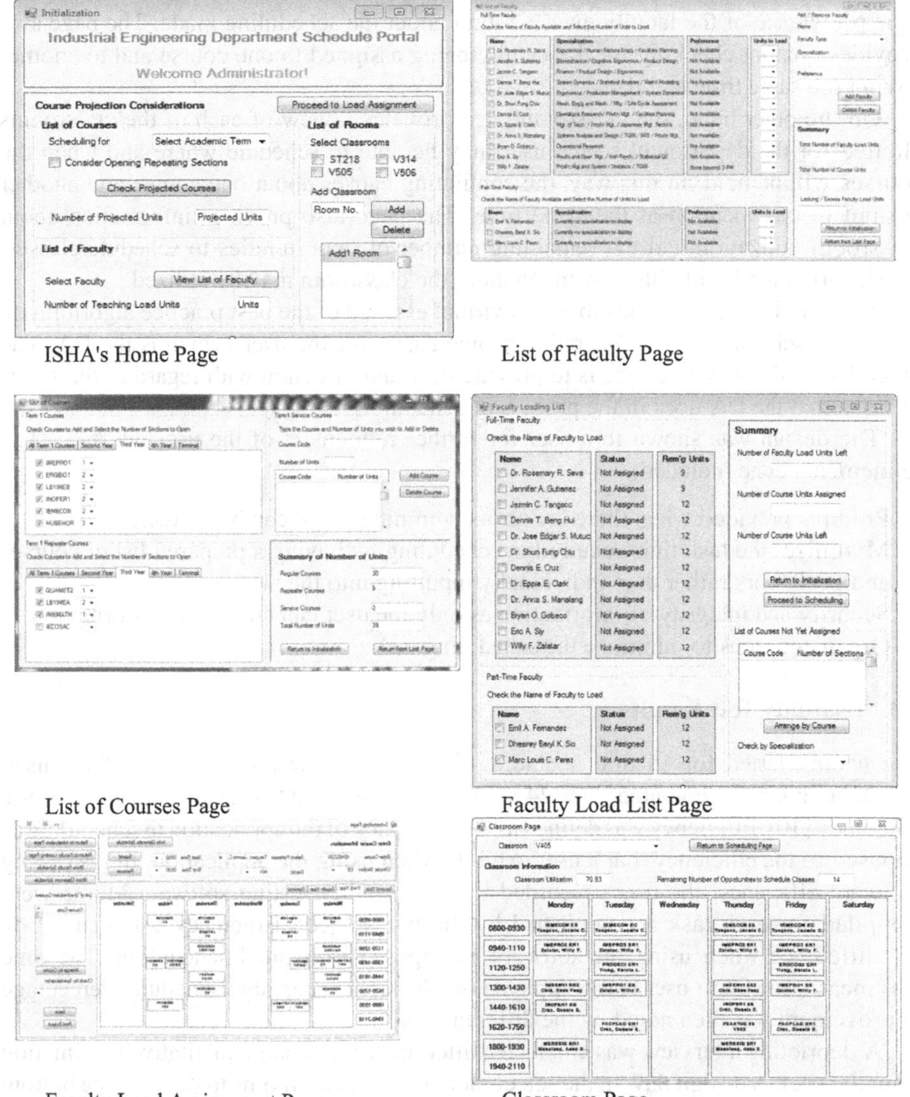

ISHA's Home Page
List of Faculty Page
List of Courses Page
Faculty Load List Page
Faculty Load Assignment Page
Classroom Page

Fig. 3. ISHA's Interface

The Faculty Load Assignment page will pop out once the name of the faculty is selected. In this view, the specialization of the faculty is considered when assigning courses. However, an exception can be made by using the checkbox provided below the name of the faculty posted on the upper left corner of the page. Through that, all courses in the list are unblocked and could be added to the faculty. When everyone in the faculty loading list has been assigned courses to teach, the scheduling can commence.

The faculty page gives a view of what the schedule of the professor would be and if his preference had been followed. The number of courses included within the range of time preference of the faculty gives a rating about the scheduling made. The page also provides conflict checking for the faculty being assigned to one course and to another having the same time and date.

On the other hand, the classroom page provides a view of each of the classrooms allotted for the department so the user may be able to schedule where and when the courses will be held. In this way, the conflicting entries about one course and another are put in one location at the same time. The page also provides information about classroom utilization and the remaining number of opportunities to schedule classes. The information highlights how much more the classroom may be utilized.

The user interface intends to give a virtual essence of the best practice algorithm on which the scheduling process could become easier for the user, which is the DA. The objective of the user interface is to provide flow and direction with regard to the series of tasks that the DA does in the process of scheduling.

The design was shown to a user for further refinement of the user interface. The refinements done included:

- Prompts provided when there are errors committed so it can be corrected
- Minimizes the task time required in scheduling by having a prepared list of courses and professors rather than individually inputting into the system
- Security and integrity of information as only the user can change the information
- Organized steps to minimize the scheduling task

3.3 Usability Test Results

The testing lasted for 65 min, including delays from queries/comments of the user. The schedule was generated after 24 s. One of the usability measures considered was efficiency [10]. Efficiency was defined as the efficiency of the application in scheduling as opposed to the efficiency that a user exhibits when using the application. In measuring software efficiency, the times recorded by the usability testing software Morae were compiled for each task accomplished by the test subject. Efficiency was gauged by the difference when using the software as opposed to manual scheduling. The time was measured after a user had completed one feasible acceptable schedule. Percentage improvement was then noted by the difference of the two times.

A debriefing interview was conducted after the test to obtain qualitative information from the user. A design flaw in the select room page causes rooms to move to the bottom of the list when they are edited. The edit buttons were also found to be too small for the user to see. In addition, when editing, the user had to click the cursor over the approximate area where the edit button was for it to appear. Additionally, the area surrounding the edit button is too small to be clicked properly. Another issue raised is the absence of a click-all button, especially in selecting professors.

Located above the web application is the progress tracker. However, another function of the progress tracker is to jump between steps or go back to previous steps quickly. However, this function was not evident when using the web application, and the progress tracker just looked like a simple progress bar with no other function.

Another test used to see if the new scheduling system had improved in terms of usability was the Systems Usability Scale survey [11]. Two survey questionnaires were given to the respondents to rate the old method and the new method of scheduling using the web app.

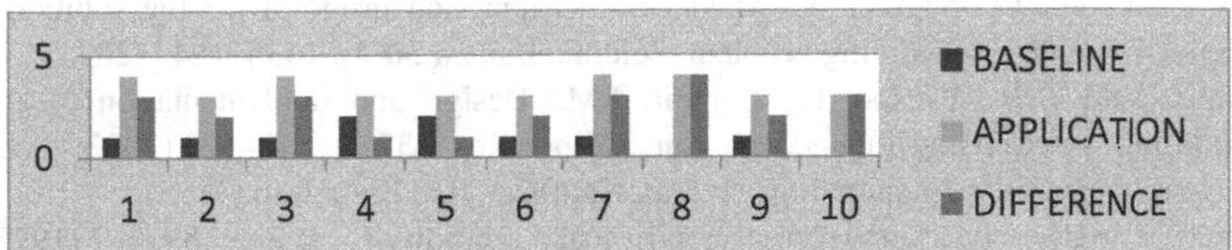

Fig. 4. SUS Survey Scores

The difference is very evident, as seen in Fig. 4. For each question, the rating for the baseline (old method), the application (new method), and the difference was charted. The overall rating of the old method had a SUS score of 25, while the application scored 85. There is an improvement of 60 SUS score units, which can be validated in the testing since the user had no errors committed.

This study has many limitations foremost of which is replicability. Since the software was developed in a specific context, only the principles of optimization can be replicated. Revisions and updates will also be required if there are major curriculum changes because the prerequisites will change. Major curriculum changes can happen every 5 years.

4 Conclusion

ISHA was able to ease the burden of the department heads in preparing a schedule that considers all the constraints imposed by the system. Two of the most important and realistic constraints solved is including the preference of the faculty in terms of time and courses which were not considered in previous studies. The scheduling task had been shortened, less prone to errors, and had been more oriented with the capabilities of the department heads. ISHA has made it possible to have the scheduling done within one (1) hour, which may only vary depending on the size of the department and the ability of the user to comprehend the system.

References

1. Bazari, S., Pooya, A., Soleimani Fard, O., Roozkhosh, P.: Modeling and solving the problem of scheduling university exams in terms of new constraints on the conflicts of professors' exams and the concurrence of exams with common questions. Opsearch **60**(2), 877–915 (2023)

2. Fedkin, E., Denissova, N., Krak, I., Dyomina, I.: Automation of scheduling training sessions in educational institutions using genetic algorithms, pp. 278–283 (2021)
3. Guia, A.D., Ballera, M.A.: Multi-agent class timetabling for higher educational institutions using prometheus platform. Indon. J. Electr. Eng. Comput. Sci. **22**(3), 1679–1687 (2021)
4. Laguardia, J.J., Flores, J.A.: University class schedule assignment by a Tabu search algorithm, pp. 728–732 (2022)
5. Wyne, M.F., Farahani, A., Atashpaz-Gargari, E., Zhang, L.: Multi-semester course staffing optimization (2022)
6. Zhang, L.H.: Research on university course timetabling problem based on divide and conquer (2022)
7. Hekmati, A., Krishnamachari, B., Matarić, M.J.: Course scheduling to minimize student wait times for university buildings during epidemics, pp. 4365–4370 (2021)
8. Can, E., Ustun, O., Saglam, S.: Metaheuristic approach proposal for the solution of the bi-objective course scheduling problem. Scientia Iranica **30**(4), 1435–1449 (2023)
9. Alvarez-Valdes, R., Crespo, E., Tamarit, J.M.: Design and implementation of a course scheduling system using Tabu search. Eur. J. Oper. Res. **137**(3), 512–523 (2002)
10. Nielsen, J.: Usability Engineering. Morgan Kaufman, San Francisco (1993)
11. Brooke, J.: SUS: a quick and dirty usability scale. Usability Eval. Ind. **189**, 4–7 (1995)

A New Data-Driven Modelling Framework for Moisture Content Prediction in Continuous Pharmaceutical Tablet Manufacturing

Motaz Deebes[1(✉)], Mahdi Mahfouf[1], and Chalak Omar[2]

[1] Department of Automatic and Control System Engineering, University of Sheffield, Sheffield, UK
{mmdeebes1,m.mahfouf}@sheffield.ac.uk

[2] Department of Multidisciplinary Engineering Education, University of Sheffield, Sheffield, UK
c.omar@sheffield.ac.uk

Abstract. Continuous manufacturing of pharmaceutical tablets integrates multiple unit operations such as twin screw granulation and fluidized bed drying to transform powder into final dosage form such as tablets. However, complex process interactions can lead to variability in critical quality attributes including moisture content of the produced granules. This study presents an innovative multi-stage modelling framework to predict granule moisture content based on the twin screw granulator and the fluidized bed dryer process parameters. Machine learning techniques, including gradient boosting regression, and support vector regression were utilised to enhance predictive performance in ensemble method. Using data from a pilot-scale integrated continuous line, the stacking ensemble model achieved excellent accuracy with (R^2) of 91% for moisture content prediction. The Machine learning modelling framework demonstrates strong potential for advancing process knowledge, and optimization in continuous manufacturing of pharmaceutical tablets based on wet granultion.

Keywords: Continuous Manufacturing · Twin Screw Granulation · Fluidized Bed Drying · Machine Learning · Pilot Plant Scale · Predictive Modelling

1 Introduction

Pharmaceutical oral solid dosages such as tablets are one of the largest industries that handle large amounts of particulate material [1]. Pharmaceutical tablet development used to rely on batch-wise manufacturing which is considered to be a consuming operation with a heavy workload and long production time [2,3]. However, continuous processing is transforming the pharmaceutical industry's

S.-H. Sheu (Ed.): ICIEA-EU 2024, LNBIP 507, pp. 107–120, 2024.
https://doi.org/10.1007/978-3-031-58113-7_10

economic and regulatory requirements, requiring compliance to speed up and improve processes [2,3]. The continuous production of pharmaceutical tablets involves feeding, wet granulating, drying, milling, and compacting [3]. All such unit operations are used sequentially to process formulation material into tablets without material feeding or removal interruptions. The pharmaceutical manufacturing sector lags in efficiency and process understanding when compared to other continuous chemical processes [4]. For example, it still faces challenges such as lack of process traceability, inconsistent product quality, and time-consuming quality assessment, which increase process failure and product waste [2,3]. Continuous processing featuring wet granulation is integral to manufacturing oral solid dosages in the pharmaceutical industry, aiding in homogeneous distribution of Active Pharmaceutical Ingredients (API) and improved particle flow [5,6]. The granulation process uses high-shear systems such as the Twin-Screw Granulator (TSG) to transport and aggregate particle material continuously while spraying binder liquid on various screw regions [7]. The TSG is commonly employed, where key parameters, including the liquid-to-solid (L/S) ratio and screw speed, influence granule properties such as size, density, and surface area [7,8]. After granulation, drying is conducted in order to eliminate moisture, commonly using Fluidized Bed Dryers (FBD) [5]. The performance of the FBD efficiency is governed by factors such as inlet temperature, drying time and air flow rate which impacts the granules' moisture content, a critical intermediate quality attribute [5,9]. This moisture, in turn, affects the tablet's physical properties, emphasizing the importance of careful monitoring and control during the drying stage for quality assurance [5,9,10]. The scope of this research is specifically focused on the granulation and drying processes in continuous pharmaceutical tablet manufacturing. Figure 1 illustrates the stages involved in the continuous manufacturing of pharmaceutical tablets via the wet granulation route.

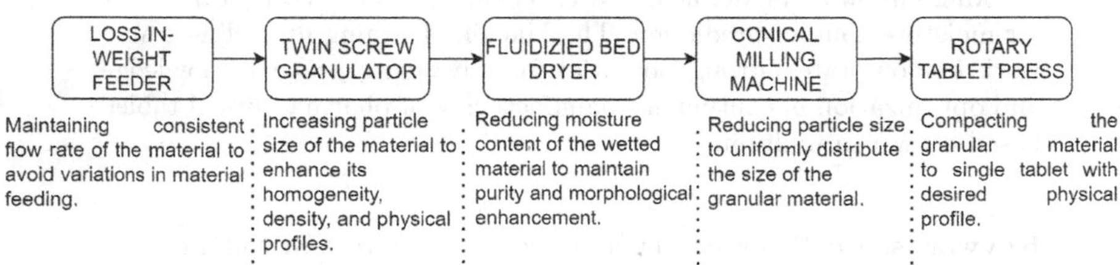

Fig. 1. Continuous Powder to Tablet Processing System block diagram.

Pharmaceutical tablet manufacturing is distinguished by complex interactions between material properties and process parameters, influenced by non-linear dynamics [4,5,11]. Predictive models, including discrete element models (DEM) [12] and population balance models (PBM) [13], have been employed to address these complexities but they are computationally inefficient [4]. Recent advances in data-driven models offer novel insights for process understanding and control [14,15]. For instance, Radial Basis Function Neural Networks (RBFNN) were developed by AlAlaween et al. to estimate granule properties in high shear granulating systems for elucidating the impact of operational parameters on material characteristics [14]. Similarly, AlAlaween et al. introduced a fuzzy logic system (FLS) optimized for twin screw granulation prediction to reveal the process complexity [15]. Other studies also investigated drying processes using a range of modelling techniques [9,16]. Monaco et al. aimed to predict optimal drying times in twin screw granulation by devising a derivative method enabling the calculation of drying rates for each specific L/S level, while Silva et al. used multivariate techniques like PCA and PLS for operational data analysis of twin-screw wet granulation and fluid bed drying which has been proven effective in detecting operational changes [16,17]. These data-driven models, while effective, face challenges as more variables are introduced, increasing model complexity and uncertainty.

The present work introduces a multi-layered, data-driven modelling framework using machine learning algorithms for accurate moisture content estimation in dried granules. The framework leverages ensemble techniques like averaging and stacking to overcome limitations associated with handling intricate, and nonlinear data. It incorporates data collection strategy from continuous powder-to-tablet pilot plant, utilizing an orthogonal array-based design of experiments to explore key parameters in interconnected processes like twin-screw granulator and fluidized bed dryer. The collected data underwent preprocessing and statistical analysis, followed by iterative stages of model training, optimization and evaluation. This approach aims to improve predictive accuracy in pharmaceutical manufacturing while adhering to Quality by Design (QbD) principles [18].

2 Methodology

The investigations presented in this work are based on the utilisation of the Consigma25 Pilot Plant (GEA Pharma Systems, Wommelgem, Belgium), available at the University of Sheffield, which involves five interconnected unit operations to process particulate material continuously. This remarkable facility, as shown in Fig. 2, includes a powder-to-tablet processing system with integrated processes that can be operated to collect operational data for different processing conditions that can be utilised for developing and validating predictive models [9,16].

2.1 Processing Materials

In this research study, a powder blend consisting of pharmaceutically relevant materials, specifically alpha-lactose monohydrate powder (Pharmatose 200M, DFE Pharma, Germany), microcrystalline cellulose powder (Pharmacel 101, DFE Pharma, Germany), and polyvinylpyrrolidone powder (Povidone K30, Huanshan Bonsun, China), has been utilised. To attain consistent and uniform mixing, the powder mixture passed through a 10-min blending process using the tumbler mixer (INVERSINA 20L, TSNS-biotech). Subsequently, the mixture was transported to the feeding unit, which includes the main hopper and the loss-in-weight feeder. This unit supplies the twin screw granulation and fluidized bed drying processes continuously with the mixture. Table 1 shows the mass fraction of each particulate material used to form the powder blend.

Fig. 2. CONSIGMA25-Continuous Powder to Tablet Processing System by GEA Pharma Systems at The University of Sheffield.

Table 1. The Powder Mixture Utilized in the Present Research Study

Material	Mass Fraction
Alpha Lactose monohydrate powder (Lactose)	72%
Microcrystalline cellulose powder (MCC)	24%
Polyvinylpyrrolidone (PVP)	4%

2.2 The Experimental Work

An efficient data collection strategy was deployed using an orthogonal array-based design of experiments [19]. Employing such approach offers the notable benefit of systematically studying multiple process variables' impacts on a response variable with minimal experiments, while generating balanced and representative data that is crucial for developing predictive models. Preliminary tests identified five key variables for each process, leading to the selection of an L27 orthogonal array that balanced complexity with efficiency of experimenting

27 runs with varying process conditions as shown in Table 2. The granulation process was conducted utilising three process parameters, namely powder feed rate, liquid feed rate, and screw speed. The powder feed rate was maintained at a constant value of 166 g/m, while the screw speed and liquid to solid ratio were manipulated across different levels, as presented in Table 2. Wet granules were transferred to a fluidized bed dryer utilising two cells, each having a filling time of 180 s. Critical drying variables—time, temperature, and inlet air flow rate—were adjusted Table 2. Moisture content was measured post-process using the near-infrared (NIR) probe [20] and the average moisture content of the two cells was recorded. This strategy identified a dataset for the key process factors, laying the foundation for data-driven models development.

Table 2. The Process Parameters Utilised for Data Collection and Process Operation

Unit Operation	Process Parameter	Parameter Level
Twin Screw Granulator	L/S Ratio	0.1, 0.2, 0.3
	Screw Speed (rpm)	400, 600, 800
Fluidized Bed Dryer	Drying Temperature (°C)	50, 60, 70
	Drying Time (s)	500, 600, 700
	Inlet Air Flow ($m^3 h^{-1}$)	300, 350, 400

2.3 Model Framework Strategy

A predictive modelling framework for both the granulation and drying processes is essential for accurate forecasting of critical process outcomes such as granule moisture content in the continuous manufacturing of pharmaceutical tablets. In this research work, the modelling framework integrates variables from both granulation (e.g., liquid/solid ratio, screw speed) and drying (e.g., drying time, drying temperature, inlet air flow) stages as input features for the model. This approach underscores the interconnected nature of the multi-stage manufacturing process, where inputs from one stage significantly affect the outcomes of the next. Thus, the predictive models were developed using two algorithms: gradient boosting regression (GBR) and support vector regression (SVR) [21–24]. GBR was utilised as an ensemble machine learning technique aimed at predicting a continuous variable through the aggregation of multiple simple models, such as decision trees [21,22]. In this method, simple models were sequentially added based on distinct sets of training data points. Residuals between estimated and actual values were calculated and then fitted into subsequent simple models with the aim of error minimization through a differentiable loss function. In addition, SVR was used as another machine learning technique for estimating continuous variables [21]. This technique aims to discover a hyperplane in a high-dimensional space that best approximates the relation between input variables and the continuous target variable within a given margin of error [23]. A polynomial kernel function was employed to map the original data into a higher-dimensional space, thereby handling non-linearity and allowing the identification of complex relationships within the data [21]. Thus, both of these learning algorithms were deemed potential candidate in recognising the process pattern for continuous processing of pharmaceutical tablets [25]. The integration of these distinct modelling techniques was undertaken to bolster both the robustness and accuracy of process predictions by combining them in an ensemble way, such as using averaging methodology and stacking methodology [21,26].

2.4 Modelling Development

In this study, the dataset derived from the experimental runs involving the twin-screw granulator and fluidized bed dryer was partitioned into input features X and target output Y. Here, X signifies operational process parameters as described in Table 2, while Y pertains to moisture content, the process output. This data was randomly divided into two subsets: 80% designated for training and the remaining 20% reserved for testing. The training set served dual purposes: initially, it facilitated model hyperparameter optimization via k-fold cross-validation, with five folds selected to ensure effective model tuning exclusively based on training data. Subsequently, this same training subset was employed to construct the final model, incorporating the optimally-tuned hyperparameters. Model efficacy was then assessed using the 20% testing subset, which had been excluded from any prior training or optimization tasks. It should be noted that prior to any model training or optimization, input features of both the training and testing sets underwent individual standardization to mitigate scale-induced bias, expedite the learning process, and prevent data leakage thereby enhancing model development performance [21]. Furthermore, the accuracy of the predictive models were evaluated using two pivotal metrics: the coefficient of determination (R^2) and the root mean squared error (RMSE) [21]. The R^2 metric quantified the extent to which the independent variables could predict the variance in the dependent variables, thereby providing an indicator of model fit relative to the actual observations [24]. In contrast, RMSE served as a measure of the discrepancies between predicted and observed values, with lower RMSE values signifying superior model fit [21,24]. Mathematically, the coefficient of determination (R^2) is defined as

$$R^2 = 1 - \frac{\sum_{i=1}^{n}(y_i - \hat{y}_i)^2}{\sum_{i=1}^{n}(y_i - \bar{y})^2} \tag{1}$$

where y_i represents the ith observed value, $\hat{y}_i$ denotes the ith predicted value, and $\bar{y}$ is the mean of the observed values. Also, the root mean squared error (RMSE) is computed as

$$\text{RMSE} = \sqrt{\frac{1}{n}\sum_{i=1}^{n}(y_i - \hat{y}_i)^2} \tag{2}$$

where y_i and $\hat{y}_i$ have the same meanings as in the R^2 equation, and n is the total number of observations. The Python programming language was primarily utilised to develop the models, hence leveraging various library resources [27–31].

Model Optimisation. Hyperparameter optimization is crucial for enhancing model performance and minimizing uncertainty [21]. In this study, individual optimization algorithms were used for each model to fine-tune specific hyperparameters while cross-validating for the lowest root mean square error (RMSE). For Gradient Boosting Regression (GBR), five key hyperparameters were adjusted: the number of estimators, the tree depth, the learning rate, the minimum number of samples, and the subsampling fraction, contributing to model accuracy [22]. Support Vector Regression (SVR) involved tuning the regularisation parameter (C), error margin (ϵ), and polynomial kernel degree to capture nonlinear patterns [21,23]. An iterative hyperparameter tuning process was conducted, targeting optimal RMSE values [32,33].

The Averaging Model Approach. In order to improve the accuracy of predictions, a potential methodology was employed that involved combining Support Vector Regression (SVR) and Gradient Boosting Regression (GBR) models through the use of averaging. The motivation behind employing this technique lies in the fundamental principle that averaging the predictions from multiple models often leads to a reduction in the variance and potentially a more robust generalization to unseen data [21]. In the present study, the models underwent training using the same set of training data, while the optimal hyperparameters were selected to maximise predictive performance. Following the completion of the training phase, the averaging procedure was performed by aggregating the predictions generated by both the Support Vector Regression (SVR) and Gradient Boosting Regression (GBR) models for each individual observation, and subsequently computing the arithmetic mean. The output is utilised for subsequent evaluation and analysis.

The Stacking Model Approach. In this study, stacking was utilized as an advanced ensemble learning methodology to harness the predictive capabilities of multiple machine learning models [26]. Unlike simpler ensemble techniques, stacking employs a two-tier architecture that includes base models and a meta-model. In our implementation, Support Vector Regression (SVR) and Gradient Boosting Regressor (GBR) were deployed as base models, while ridge regression served as the meta-model [25,26]. This meta-model acts as an integrator of these predictions. It assigns coefficient weights to the predictions from the base models, effectively learning how to optimize and balance their contributions. The utilization of Ridge regression is strategic; its regularization term prevents the meta-model from overfitting among predictions from the base models, thereby enhancing the ensemble's generalizability and predictive power [21].

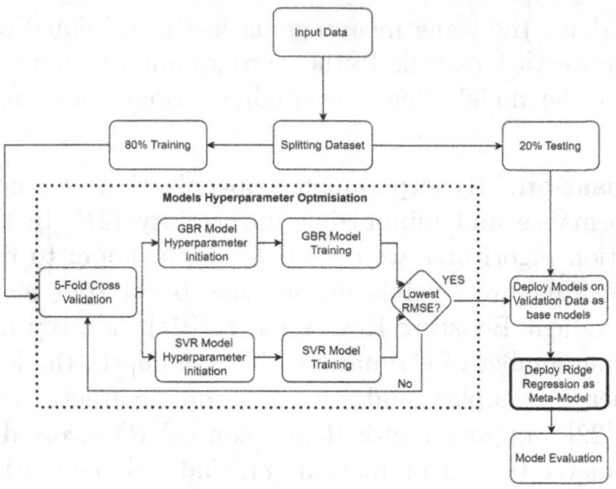

Fig. 3. Stacking Model Flow chart. Here the predicted data are fitted to a meta-model in this case Ridge Regression model

Comprehensive evaluation metrics were utilised to assess the stacking model's performance. Figure 3 is the block diagram showing the model development strategy for such approach.

3 Results and Discussions

Operational excellence in the continuous manufacturing of pharmaceutical tablets relies on the optimisation and control of multiple unit operations. Specifically, the utilisation of Twin-Screw Granulation (TSG) and Fluidized Bed Drying (FBD) processes is of utmost importance as they have a substantial influence on the quality of the end product [9,14]. This section includes the statistical analysis of data collected from the continuous processes of twin-screw granulation and fluidized bed drying. Additionally, it evaluates the model performance to underscore the efficacy of the proposed approaches in capturing the complexities inherent in continuous pharmaceutical tablet manufacturing.

3.1 Experimental Analysis

In this study, a comprehensive statistical analysis, including Pearson-based bivariate correlation, was conducted to assess the relationship between key process variables—such as L/S ratio, screw speed, and drying conditions—and material moisture content. The coefficient r is a measure that ranges from -1 to 1, indicating the strength and direction of the linear relationship between two variables [24]. The L/S ratio showed a strong positive correlation $r = 0.8$ with moisture content, while screw speed had a low correlation $r = 0.1$. Drying-related variables such as drying time, temperature, and inlet air flow displayed moderate negative correlations with moisture content ($r = -0.1, -0.3, -0.2$ respectively).

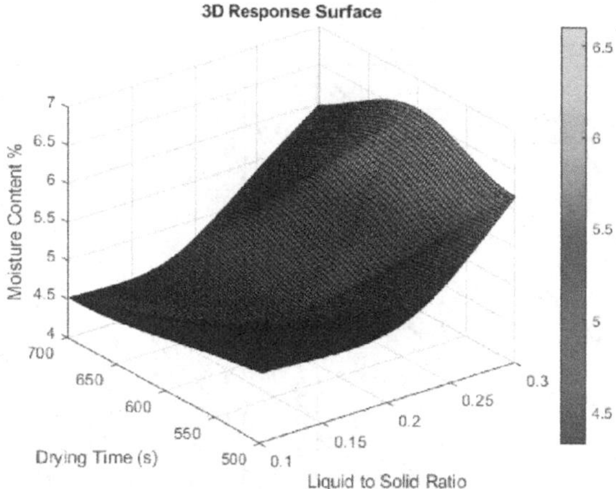

Fig. 4. Response surface showing the effect of drying time and liquid to solid ratio on moisture content. Extended drying times and lower liquid-to-solid ratios are associated with reduced moisture content.

In addition, a 3D response surface plot is implemented to gain insights into how varying processing parameters impact the moisture content, as depicted in Fig. 4. Thus, this 3D plot and the correlations, depicted in Fig. 5, imply the possibility of non-linear relationships, warranting advanced modelling techniques for a full understanding.

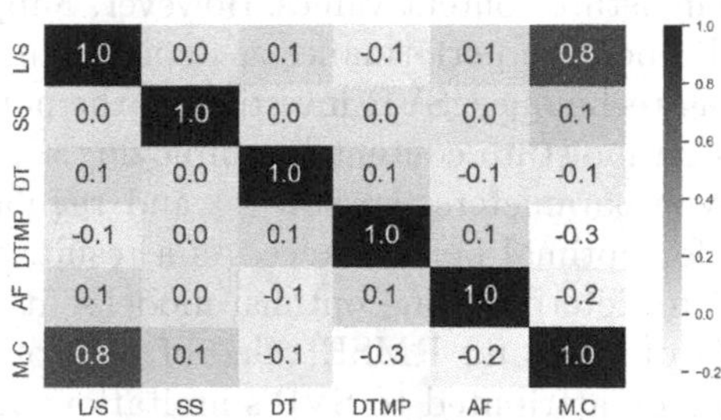

Fig. 5. Correlation map between the process variables and the process outcome. Here L/S: Liquid to Solid ratio, SS: the Screws Speed(rpm), DT: Drying Time (s), DTMP: Drying Temperature (°C), AF: Inlet Airflow Rate (m^3/h), M.C: Moisture Content (w/w)

3.2 Performance of Individual Models

In this research, GBR was particularly effective in predicting the moisture content based on the process conditions of twin screw granulation and fludizied bed drying. It has demonstrated notable strength to handle limited and complex data, which are commonplace in pharmaceutical processes, achieving (R^2)

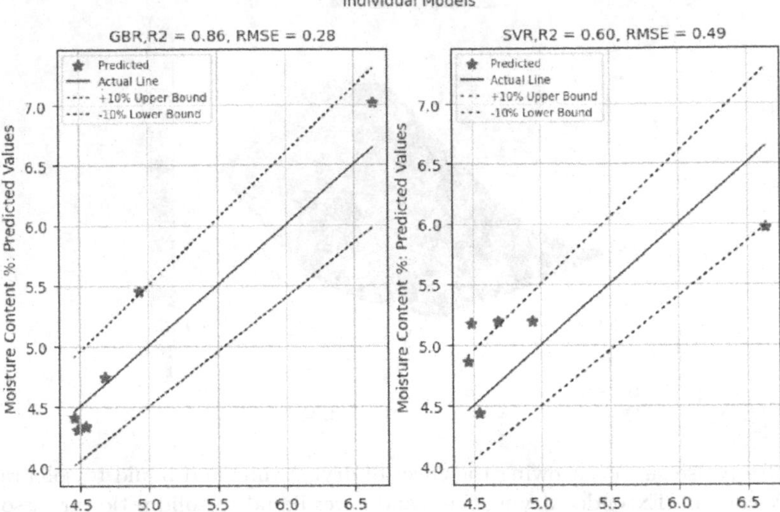

Fig. 6. Comparison of GBR and SVR models performance on unseen data.

of 86% and RMSE value of 0.28 on unseen data, suggesting that it captured a significant portion of the variance in the data. This is due to the fact that the GBR is handling uncertainty as part of its learning process by rectifying errors iteratively. As depicted in Fig. 6, the plot confirms a close correlation between predicted and actual moisture content values. However, Support Vector Regression (SVR) displayed moderate performance as depicted in Fig. 6. Utilising the mathematical mapping techniques, SVR investigated the non-linear interactions of process conditions on moisture content by employing a polynomial kernel of degree three. Key hyperparameters, epsilon (ϵ) and regularization parameter (C), were fine-tuned for optimal performance. As a result, a total of 8 support vectors were selected to determine the optimal model's fit. It achieved a comparatively lower (R^2) of 0.60 and RMSE value of 0.49 on unseen data. This moderate efficacy may be attributed to SVR's limitations in handling complex data. Despite these shortcomings, the algorithm still managed to capture essential aspects of the data, albeit with slight inconsistencies, as seen in Fig. 6.

3.3 Performance of Averaging Model

The averaging model serves as a basic yet efficient ensemble method in which the predictions from support vector regression (SVR) and gradient boosting regression (GBR) were averaged to provide enhanced predictive performance. Thus, by averaging their predictions, the coefficient of determination (R^2) of 89% is obtained, while the root mean square error RMSE was 0.25 reflecting a significant performance of such modelling approach. In fact, when these two models are combined by averaging their predictions, they encapsulate a more comprehensive understanding of the data. For example, the averaging model's performance is remarkably close to the GBR model's with (R^2) of 86% but with a better generalization. Furthermore, the stability of the averaging model, in which the models' predictions help in muting the sensitivities and limitations of each model, rendering a more stable and robust prediction. As depicted in Fig. 7, a plot of the actual versus averaged predicted values revealed a good alignment along the actual line compared to the individual models. While offering improvements in predictive accuracy, the averaging model's key advantage lies in its simplicity, which translates into ease of interpretation.

3.4 Performance of Stacking Model

The stacking model, comprising Support Vector Regressor (SVR) and Gradient Boosting Regressor (GBR) as base models, and Ridge Regression as the meta-model, demonstrated marked improvements in predictive accuracy for continuous pharmaceutical tablet manufacturing. This model achieved an RMSE value of 0.23 and an (R^2) value of 91%, outperforming all other models discussed previously. Notably, Ridge Regression served a pivotal role in this enhanced performance. The Ridge Regression meta-model is designed to intelligently weight and combine the predictions of the base models, thereby maximizing predictive accuracy. Its inclusion of $L2$ regularization term enables fine-tuning of prediction uncertainties from the base models, preventing overfitting and offering a more robust model. This regularization feature contributed to a slight improvement over the averaging model. Figure 8 provides a visual representation of these results, showcasing a plot of actual versus predicted values. The data points are closely aligned along the actual line of strong agreement, affirming the model's ability to synthesize information from multiple base models without introducing bias or overfitting to the training data. Moreover, the stacking approach has succeeded in capturing the data pattern within a 10% error band suggesting the potential use of such approach for further predictive modelling for the continuous manufacturing of pharmaceutical tablets.

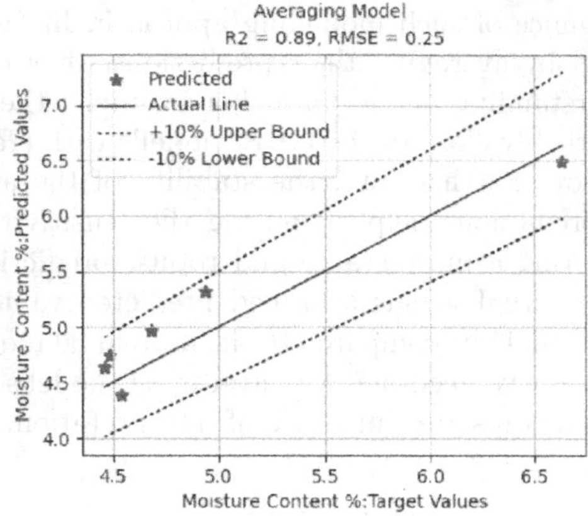

Fig. 7. Performance of Averaging Model based on testing data

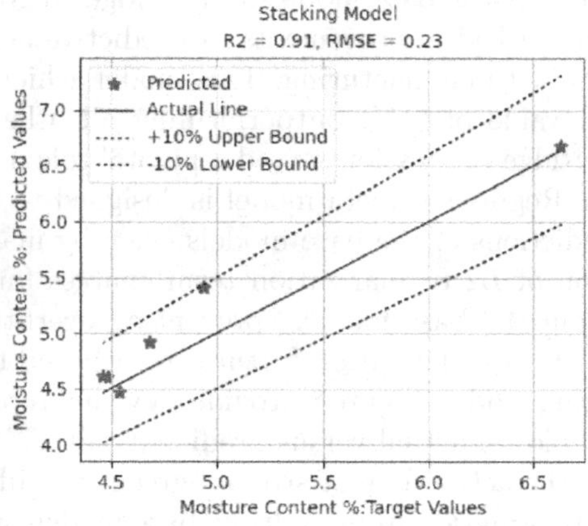

Fig. 8. Performance of Stacking Model based on testing data

4 Conclusion

In summary, this study proposed a multi-stage modelling framework using machine learning methods such as gradient boosting and support vector regression, along with ensemble techniques including averaging and stacking, to predict moisture content in a continuous pharmaceutical manufacturing process. Using orthogonal design of experiments, critical variables in twin-screw granulation and fluidized bed drying were investigated. The stacked ensemble model, featuring SVR and GBR as base models and ridge regression as the meta-model, yielded superior performance (R^2) value of 91% and RMSE of 0.23. The framework successfully mapped complex process-moisture relationships, aligning with Quality by Design principles, and has broad implications for process optimization and control. In future research, this modelling framework will evolve into a sequential, plant-wide paradigm for optimizing the continuous manufacturing of pharmaceutical tablets. Ultimately, this leads to a holistic understanding and improved key quality attributes, in line with a 'right-first-time' approach.

Acknowledgements. The authors wish to thank DFE Pharmaceuticals for providing the alpha-lactose monohydrate powder (Pharmatose 200M), and the microcrystalline cellulose powder (Pharmacel 101), used in this study. DFE Pharmaceuticals generous contribution of these pharmaceutically relevant materials was instrumental in enabling the experimental work and data collection that made this research possible.

References

1. Litster, J.: Design and Processing of Particulate Products. Cambridge University Press, Cambridge (2016)
2. Lee, S.L., et al.: Modernizing pharmaceutical manufacturing: from batch to continuous production. J. Pharm. Innov. **10**, 191–199 (2015)
3. Khinast, J., Bresciani, M.: Continuous Manufacturing: Definitions and Engineering Principles, chsp. 1, pp. 1–31. Wiley (2017)
4. Rogers, A.J., Hashemi, A., Ierapetritou, M.G.: Modeling of particulate processes for the continuous manufacture of solid-based pharmaceutical dosage forms. Processes **1**, 67–127 (2013)
5. Sacher, S., Khinast, J.G.: An overview of pharmaceutical manufacturing for solid dosage forms. In: Ierapetritou, M.G., Ramachandran, R. (eds.) Process Simulation and Data Modeling in Solid Oral Drug Development and Manufacture. MPT, pp. 311–383. Springer, New York (2016). https://doi.org/10.1007/978-1-4939-2996-2_10
6. Litster, J., Ennis, B.: The Science and Engineering of Granulation Processes, vol. 15 (2004)
7. Seem, T.C., et al.: Twin screw granulation - a literature review. Powder Technol. **276**, 89–102 (2015)
8. Vercruysse, J., et al.: Continuous twin screw granulation: influence of process variables on granule and tablet quality. Eur. J. Pharm. Biopharm. **82**, 205–211 (2012)
9. Monaco, D., Omar, C., Reynolds, G.K., Tajarobi, P., Litster, J.D., Salman, A.D.: Drying in a continuous wet granulation line: investigation of different end of drying control methods. Powder Technol. **392**, 157–166 (2021)
10. Thapa, P., Lee, A.R., Choi, D.H., Jeong, S.H.: Effects of moisture content and compression pressure of various deforming granules on the physical properties of tablets. Powder Technol. **310**, 92–102 (2017)
11. Boukouvala, F., Muzzio, F.J., Ierapetritou, M.G.: Dynamic data-driven modeling of pharmaceutical processes. Industr. Eng. Chem. Rese. **50**, 6743–6754 (2011)
12. Ketterhagen, W.R., Ende, M.T.A., Hancock, B.C.: Process modeling in the pharmaceutical industry using the discrete element method. J. Pharm. Sci. **98**, 442–470 (2009)
13. Chaudhury, A., Sen, M., Barrasso, D., Ramachandran, R.: Population balance models for pharmaceutical processes. In: Ierapetritou, M.G., Ramachandran, R. (eds.) Process Simulation and Data Modeling in Solid Oral Drug Development and Manufacture. MPT, pp. 43–83. Springer, New York (2016). https://doi.org/10.1007/978-1-4939-2996-2_2

14. AlAlaween, W.H., Mahfouf, M., Salman, A.D.: Predictive modelling of the granulation process using a systems-engineering approach. Powder Technol. **302**, 265–274 (2016)
15. Wafa'H, A., Khorsheed, B., Mahfouf, M., Reynolds, G.K., Salman, A.D.: An interpretable fuzzy logic based data-driven model for the twin screw granulation process. Powder Technol. **364**, 135–144 (2020)
16. Silva, A.F., et al.: Process monitoring and evaluation of a continuous pharmaceutical twin-screw granulation and drying process using multivariate data analysis. Eur. J. Pharm. Biopharm. **128**, 36–47 (2018)
17. Kourti, T.: Process analytical technology beyond real-time analyzers: the role of multivariate analysis. Crit. Rev. Anal. Chem. **36**, 257–278 (2006)
18. Yu, L.X., Amidon, G., Khan, M.A., Hoag, S.W., Polli, J., Raju, G.K., Woodcock, J.: Understanding pharmaceutical quality by design. AAPS J **16**, 771–783 (2014)
19. Hedayat, A.S., Sloane, N.J.A., Stufken, J.: Statistical application of orthogonal arrays. In: Hedayat, A.S., Sloane, N.J.A., Stufken, J. (eds.) Orthogonal Arrays. Springer Series in Statistics, pp. 247–315. Springer, New York (1999). https://doi.org/10.1007/978-1-4612-1478-6_11
20. Romanãch, R.J., Román-Ospino, A.D., Alcalà, M.: A procedure for developing quantitative near infrared (NIR) methods for pharmaceutical products. Methods Pharmacol. Toxicol. **32**, 133–158 (2016)
21. Kuhn, M., Johnson, K.: Applied Predictive Modeling. Springer, New York (2013). https://doi.org/10.1007/978-1-4614-6849-3
22. Natekin, A., Knoll, A.: Gradient boosting machines, a tutorial. Front. Neurorobot. **7**, 21 (2013)
23. Drucker, H., Burges, C.J.C., Kaufman, L., Smola, A., Vapnik, V.: Support vector regression machines. In: Mozer, M., Jordan, M., Petsche, T. (eds.) Advances in Neural Information Processing Systems, vol. 9. MIT Press (1996)
24. James, G., Witten, D., Hastie, T., Tibshirani, R.: An Introduction to Statistical Learning: with Applications in R. Springer, New York (2013). https://doi.org/10.1007/978-1-4614-7138-7
25. Lou, H., Lian, B., Hageman, M.J.: Applications of machine learning in solid oral dosage form development. J. Pharm. Sci. **110**, 3150–3165 (2021)
26. Wolpert, D.H.: Stacked generalization. Neural Netw. **5**(2), 241–259 (1992)
27. McKinney, W.: Data structures for statistical computing in python. In: van der Walt, S., Millman, J. (eds.) Proceedings of the 9th Python in Science Conference, pp. 56–61 (2010)
28. Pedregosa, F., et al.: Scikit-learn: machine learning in Python. J. Mach. Learn. Res. **12**, 2825–2830 (2011)
29. Harris, C.R., et al.: Array programming with NumPy. Nature **585**(7825), 357–362 (2020)
30. Waskom, M.L.: seaborn: statistical data visualization. J. Open Sour. Softw. **6**(60), 3021 (2021)
31. Hunter, J.D.: Matplotlib: a 2D graphics environment. Comput. Sci. Eng. **9**(3), 90–95 (2007)
32. Bergstra, J., Bardenet, R., Bengio, Y., Kégl, B.: Algorithms for hyper-parameter optimization. In: Advances in Neural Information Processing Systems, vol. 24 (2011)
33. Akiba, T., Sano, S., Yanase, T., Ohta, T., Koyama, M.: Optuna: a next-generation hyperparameter optimization framework. In: Proceedings of the 25th ACM SIGKDD International Conference on Knowledge Discovery and Data Mining (2019)

Solving the Car Sequencing Problem with Cross-Ratio Constraints Using Constraint Programming Approach

Sana Jalilvand[1], Ali Bozorgi-Amiri[2], Mehdi Mamoodjanloo[1], and Armand Baboli[1](✉)

[1] LIRIS Laboratory, UMR 5205 CNRS, INSA of Lyon, Villeurbanne, France
{sana.jalilvand,mehdi.mahmoodjanloo,armand.baboli}@insa-lyon.fr
[2] School of Industrial Engineering, College of Engineering, University of Tehran, Tehran, Iran
alibozorgi@ut.ac.ir

Abstract. The rise of mass-individualization has underscored the significance of Mixed-Model Assembly Lines (MMALs) for producing diverse products on the same line. The Car Sequencing Problem (CSP) tackles short-term balancing in an MMAL by emphasizing the use of spacing rules to manage the space between each pair of work-intensive products that possess specific characteristics. In this study, we tackle two challenges within the CSP context. The first challenge involves exploring CSP with cross-ratio constraints that takes into account the dependency between different characteristics. As the second challenge, we study the CSP under two states where spacing rule violations are not allowed (hard) and allowed (soft). We develop two constraint programming models for the mentioned states and evaluate the performance of the models using several real-world assembly lines' instances. The findings enhance understanding of each model's strengths and weaknesses. Given the inherent complexity of real-world problems, the soft model may find more practical and effective application. This research enriches the realm of problem-solving in MMALs by offering valuable insights and introducing the main challenges in the CSP.

Keywords: Car sequencing problem · Mixed-model assembly line · Constraint programming · Constraint dependency

1 Introduction and Literature Review

In recent times, with the increasing emphasis on consumer preferences for customization, the concept of mass individualization has gained prominence. This shift has resulted in a transition from traditional make-to-stock strategies to more agile make-to-order approaches. To maintain the advantages of an efficient flow production while accommodating a wide range of diverse products, Mixed-Model Assembly Lines (MMALs), can be effectively employed within this context. The MMAL allows for the assembly of multiple product models with different options/characteristics on the same line [1, 2].

S.-H. Sheu (Ed.): ICIEA-EU 2024, LNBIP 507, pp. 121–132, 2024.
https://doi.org/10.1007/978-3-031-58113-7_11

In an MMAL, the installation time of each characteristic may vary in different products. Imagine a car assembly line where cars with three kinds of electrical, CNG, and diesel engines can be ordered by customers. The cars that require electrical engines would have more complex electrical characteristics and thus impose more workload on the stations that operate these features. Hence, one of the most challenging issues in MMAL occurs when work-intensive products in the same station, follow each other leading to work overload. Work overload can disrupt production flow and hinder efficiency. To address this, a short-term balancing policy is required to develop an appropriate sequence that avoids work overload and ensures a smooth production flow.

The concept of work overload is illustrated in Fig. 1, where cars with different characteristics are sequenced based on their operation times. To avoid work overload, spacing between work-intensive cars must be set to ensure that their average cumulative operation times do not exceed the takt time. Typically, the sequencing problem should be assessed explicitly based on operation times while taking the constraints of the workstations and the logistics systems into account. However, in complex real-world scenarios involving intricate products with multiple components, an implicit sequencing strategy, known as the Car Sequencing Problem (CSP), aims to achieve short-term balancing without detailed operation allocation. CSP regulates work overload by controlling the distribution of work-intensive product characteristics. This involves compiling a list of required characteristics for each product and defining spacing rules, denoted as $H_c : N_c$. These rules stipulate that within subsequences of N products, a maximum of H products with characteristic c should be present to respect the spacing rule. These independent rules prevent consecutive work-intensive products in the sequence. Adhering to these rules yields a product sequence that mitigates work overload [3].

The recent increase in the variety of products not only impacts the mainline but also introduces restrictions in sublines or logistic systems, potentially leading to dependency between two characteristics. Such dependencies arise when two characteristics that are both work-intensive at the same station, share material handling systems or tools, or require shared preparation areas for the parts they assemble. To address this issue, it becomes essential to establish dependent spacing rules alongside the existing independent ones. These rules serve as constraints within the context of CSP. In this study, the terms "ratio" and "cross-ratio" constraints denote the independent and dependent spacing rules, respectively. The expanded version of the standard CSP that incorporates both ratio and cross-ratio constraints has recently been developed by [4] and is termed CSP-CR.

Initially introduced as a constraint satisfaction problem by [5], CSP used to belong to the category of feasibility problems rather than optimization problems. In this approach, to establish a feasible solution, all spacing rules are treated as hard constraints, aiming to identify a sequence where no rules are violated. However, due to the real-world problems' complexity, it might be challenging or even impossible to find a feasible solution in some cases. Consequently, CSP evolved into a constraint optimization problem by incorporating an objective function that tries to minimize the spacing rule violations [6–8]. This consideration turns spacing rules into soft constraints, enabling the derivation of a sequence that aids managers in identifying the stations where work overload occurs so that they can better manage the violations.

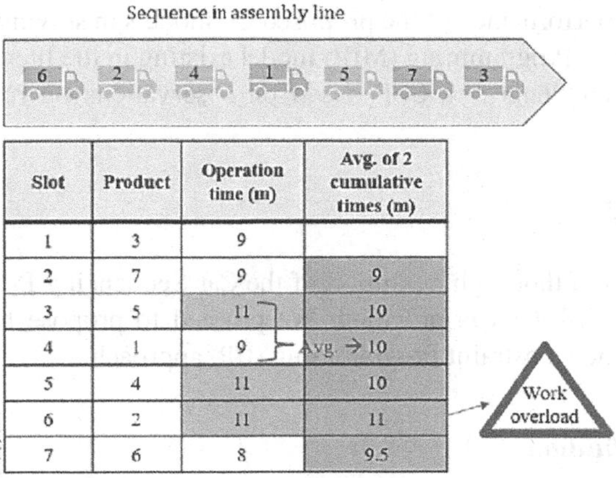

Sequence in assembly line

6 2 4 1 5 7 3

Slot	Product	Operation time (m)	Avg. of 2 cumulative times (m)
1	3	9	
2	7	9	9
3	5	11	10
4	1	9	Avg. → 10
5	4	11	10
6	2	11	11
7	6	8	9.5

Work overload

Fig. 1. A random sequence in an MMAL with work overload

CSP is addressed by both Constraint Programming (CP) and Mixed Integer Programming (MIP) approaches. These two approaches are employed to obtain an exact solution for small- to medium-sized instances. When dealing with highly constrained and complicated optimization problems, such as scheduling problems, CP can compete with MIP [9]. [10] conducted a study of CSPLib's car sequencing instances, encoded in the form of an n-ary constraint satisfaction problem. They implemented seven different value ordering algorithms based on both fail-first and succeed-first principles and reported that the former was more efficient at solving a substantial number of these instances. [11] classified a set of CP heuristics for the CSP based on 4 criteria, namely branching variables, exploration directions, parameters for the selection of branching variables, and aggregation functions for these criteria. They analyzed and compared the functionality of the proposed algorithms in solving various instances. To the best of our knowledge, CP has mostly been applied to problems with only hard constraints and without an objective function. The result of these types of problems is either a sequence that satisfies all the spacing rules or a failure indicating that there is no such feasible solution [12]. On the other hand, CSP is proven to be NP-hard in the strong sense [13]. Hence, several articles have tackled CSPs in the large scale by proposing heuristic and metaheuristic algorithms such as ant colony optimization [14], genetic algorithm [15], simulated annealing [16], variable neighborhood search [17], tabu search [18], large neighborhood search [19], etc. The objective of these studies is to identify a solution approach capable of yielding a high-quality solution within a reasonable timeframe.

The main contributions of this study are outlined as follows:

- Addressing the CSP with both independent spacing rules (ratio constraints) and dependent spacing rules (cross-ratio constraints) as a newly introduced variant of CSP,
- Studying the problem in two states: one in which the spacing rule violations are not allowed (hard constraints), and the other in which they are allowed (soft constraints) using Constraint Programming (CP),

- Comparing the performance of the proposed CP models in solving the problem with the Mixed-Integer Programming (MIP) model existing in the literature [4],
- Analyzing the challenges of CSP based on observations on real-world car final assembly lines.

2 Methodology

In this section, first, a thorough definition of the Car Sequencing Problem with Cross-Ratio constraints (CSP-CR) is provided. We proceed to propose two models of the problem utilizing the Constraint Programming (CP) approach.

2.1 Problem Definition

CSP-CR involves finding a sequence for products with different characteristics to enter the assembly line, while adhering to predefined spacing rules. In this constraint satisfaction problem, products are treated as variables, and each product's possible slot on the assembly line is defined as the domain. One constraint ensures that each slot can only be assigned to one product. Spacing rules give rise to two types of constraints: ratio constraints and cross-ratio constraints. Ratio constraints or independent rules, represented as $H_c : N_c$, limit the placement of work-intensive products in subsequences. Cross-ratio constraints stem from dependent spacing rules and are specified as $1 : N'_{c_1c_2}$. In Fig. 2, an illustration of a sequence for 6 products with three characteristics (c_1-c_3) and their corresponding rules is shown. Violations occur when spacing rules are not respected. For example, for characteristic c_3, the independent rule $1 : 3$ means that in every subsequence of 3 products, there should be a maximum of 1 product with c_3. The yellow cell in slot 3 signifies a violation of the independent rule $1 : 3$ for characteristic c_3. It means there are more than 1 product with c_3 in a subsequence of 3 products. On the other hand, if characteristic c_1 is dependent on c_2 with the rule $1 : 2$ ($N'_{c_1c_2} = 2$), there should be no product with c_1 in the next 1 (2-1) slot after a product with c_2. The orange cell represents a violation of the dependent rule1:2 between c_1 and c_2.

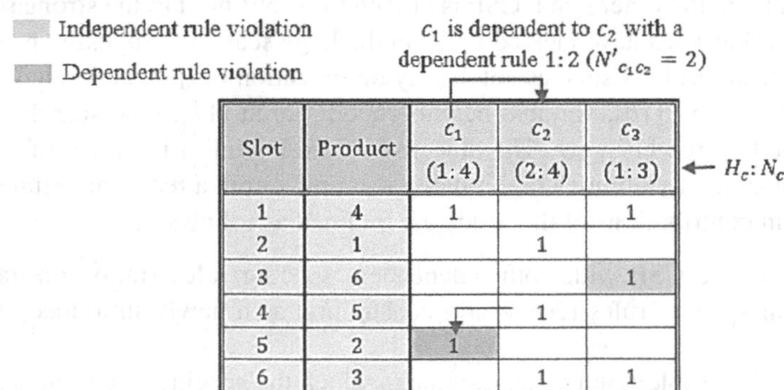

Slot	Product	c_1 (1:4)	c_2 (2:4)	c_3 (1:3)
1	4	1		1
2	1		1	
3	6			1
4	5		1	
5	2	1		
6	3		1	1

Fig. 2. A random sequence of six products with dependent and independent violations

Since CSP-CR is a type of constraint satisfaction problem, the solution of the problem (the sequence of products) is obtained when all the above-mentioned constraints are satisfied. In most real-world cases, no feasible solutions can be found if we consider all the constraints as "hard" ones, i.e., no violations occur. However, an alternative approach is to characterize constraints as "soft", count the spacing rule violations, and ultimately minimize the weighted penalty associated with these spacing rule violations. In other words, the constraint satisfaction problem with hard constraints and no objective function can be transformed into a constrained optimization problem with soft constraints and an objective function aimed at minimizing the penalties associated with violated spacing rules.

2.2 Constraint Programming (CP) Approach

Constraint programming is an approach to address a wide range of constraint satisfaction problems including scheduling, sequencing, timetabling, etc. It is a method for finding the solutions that satisfy a set of specified constraints. This approach operates by leveraging the constraints to narrow down the possible values that can be assigned to each variable, ultimately leading to a solution. CP is applied in two distinct manners:

1. Feasibility: The aim is to identify one or more feasible solutions (i.e., solutions that respect all the constraints). In this respect, we propose the CP model for CSP-CR with hard constraints in subsection "Model CP1: CSP-CR with Hard Constraints". There is no objective function in this model, and a solution is only obtained if all the ratio and cross-ratio constraints are satisfied
2. Optimization: The aim of optimization is to identify the best feasible solution given an objective function. In subsection "Model CP2: CSP-CR with Soft Constraints", a CP model of CSP-CR with soft constraints is provided. In this scenario, the problem's space is transformed into a soft state by quantifying the spacing rules violations in constraints and minimizing these violations through a penalty-based objective function.

According to the formulation of CSP-CR constraint programming model, the notations are introduced as follows:

Sets and indices:	
P	Set of products with index $p \in P$
C	Set of characteristics with index $c \in C$
S	Set of slots with index $s \in S = \{1, 2, \ldots, \|P\|\}$
Parameters:	
α_{pc}	$\begin{cases} 1, & \text{if product } p \text{ contains characterisic } c \\ 0, & \text{otherwise} \end{cases}$
$H_c : N_c$	Independent rule of characteristic c; In every subsequence of N products, there should be a maximum of H products containing characteristic c; $H_c < N_c$
w_c	Penalty weight of independent rule violation corresponding to characteristic c

(continued)

(*continued*)

$\beta_{c_1c_2}$	$\begin{cases} 1, & \text{if characterisic } c_1 \text{ is dependent to characterisic } c_2 \\ 0, & \text{otherwise} \end{cases}$
$1 : N'_{c_1c_2}$	Dependent rule between characteristics c_1 and c_2; A product containing c_1 should not come at least in $(N'_{c_1c_2} - 1)$ slots after a product containing c_2
$w'_{c_1c_2}$	Penalty weight of dependent rule violation corresponding to characteristics c_1 and c_2
M	A large positive number
Decision variables:	
$Prod_s$	Product positioned in slot s; $Prod_s \in \{1, 2, \ldots, \|P\|\}$
$Setup_{sc}$	$\begin{cases} 1, & \text{if slot } s \text{ is assigned to a product containing characterisic } c \\ 0, & \textit{otherwise} \end{cases}$
IV_{sc}	Number of independent violations of characteristic c of the product positioned in slot s
$DV_{sc_1c_2}$	Number of dependent violations for a product containing characteristic c_2 when a product containing characteristic c_1 is positioned in slot s

Model CP1: CSP-CR with Hard Constraints. Model CP1, as outlined below, represents the CP formulation of the CSP-CR with hard constraints; i.e., without allowing the occurrence of any violation. This model is specifically oriented toward the discovery of a feasible solution that adheres to all the ratio and cross-ratio constraints.

Subject to:

$$allDifferent(Prod_s) \tag{1}$$

$$Setup_{sc} = \alpha_{Prod_s,c}; \qquad \forall s \in S, c \in C \tag{2}$$

$$Setup_{sc} \times \sum_{s'=s+1}^{min\{s+N_c-1,|P|\}} Setup_{s'c} \le H_c - 1;$$
$$\forall s = 1, 2, \ldots, |P| - 1, c \in C \tag{3}$$

$$Setup_{sc_2} \times \beta_{c_1c_2} \times \sum_{s'=s+1}^{min\left\{s+N'_{c_1c_2}-1,|P|\right\}} Setup_{s'c_1} \le 0;$$
$$\forall s = 1, 2, \ldots, |P| - 1, c_1, c_2 \in C, c_1 \ne c_2 \tag{4}$$

$$Prod_s \in \{1, 2, \ldots, |P|\}; \qquad \forall s \in S \tag{5}$$

$$Setup_{sc} \in \{0, 1\}; \qquad \forall s \in S, c \in C \tag{6}$$

Constraint (1) consists of *allDifferent* function which is a specialized function in CP language. It ensures that all products take different integer values as slots. Constraint (2) creates the *Setup* binary matrix. It shows that which characteristics need to be installed in each slot, after the sequence is identified. Constraint (3) or ratio constraint is associated with the characteristics' independent spacing rules. It states that in every subsequence of N products, there should be a maximum of H products containing characteristic c. Constraint (4) or cross-ratio constraint is related to the dependent spacing rule between characteristics c_1 and c_2. It declares that if c_1 is dependent to c_2, a product containing c_1 should not appear at least in $(N'_{c_1c_2} - 1)$ slots after a product containing c_2. Constraints (5) and (6) define the domains of the decision variables.

Model CP2: CSP-CR with Soft Constraints. Model CP2 represents the CP formulation of the CSP-CR with soft constraints. In this model, in contrast to the model CP1, the rule violations are allowed, and they are minimized with a penalty weight in the objective function.

$$\min Z = \sum_{c \in C} \sum_{s=1}^{|P|} w_c.IV_{sc} + \sum_{c_1 \in C} \sum_{\substack{c_2 \in C \\ c_1 \neq c_2}} \sum_{c=1}^{|P|} w'_{c_1c_2}.DV_{sc_1c_2} \quad (7)$$

Subject to (1), (2), (5), (6), and:

$$Setup_{sc} \times \sum_{s'=s+1}^{min\{s+N_c-1,|P|\}} Setup_{s'c} \leq H_c - 1 + IV_{sc};$$
$$\forall s = 1, 2, \ldots, |P| - 1, c \in C \quad (8)$$

$$Setup_{sc_2} \times \beta_{c_1c_2} \times \sum_{s'=s+1}^{min\left\{s+N'_{c_1c_2}-1,|P|\right\}} Setup_{s'c_1} \leq DV_{sc_1c_2};$$
$$\forall s = 1, 2, \ldots, |P| - 1, c_1, c_2 \in C, c_1 \neq c_2 \quad (9)$$

$$IV_{sc} \geq 0, int; \qquad \forall s \in S, c \in C \quad (10)$$

$$DV_{sc_1c_2} \geq 0, int; \qquad \forall s \in S, c_1, c_2 \in C \quad (11)$$

Objective function (7) minimizes the total penalty associated with the both ratio and cross-ratio constraints violations. Constraints (1), (2), (5) and (6) are the same as the model CP1. Constraint (8) counts the number of independent spacing rule violations. Constraint (9) quantifies the dependent spacing rule violation occurrences. Constraints (10) and (11) define the domains of the decision variables.

3 Numerical Experiments

In this section, several numerical experiments are conducted on various instances in order to assess and compare the performance of the two models provided in Sect. 2 (CP1 and CP2). In order to validate the proposed CP models, all the instances are also run by the CSP-CR's Mixed Integer Programming (MIP) model as introduced by [4]. IBM ILOG CPLEX Optimization Studio (version 22.1.0) and GAMS (version 24.1.3) with CPLEX solver are employed to solve the instances with CP and MIP approaches, respectively. All the computations are executed by a system with processor "Intel (R) Core (TM) i7-8550U @ 1.80 GHz CPU" and RAM of "16 GB".

3.1 Case Study Problems

The computational experiments are conducted on instances extracted from three real-world car final assembly lines. Table 1 presents the details of each instance. The term UR refers to the Utilization Rate which indicates the complexity of the problem to be solved. UR is calculated by Eq. (12) [6]. Instance complexity is influenced by multiple factors, including the number of products ($|P|$), the number of products requiring each characteristic ($\sum_p \alpha_{pc_i}$), and the spacing rule's strictness ($H_{c_i} : N_{c_i}$).

$$UR(c_i) = \frac{N_{c_i}.\sum_p \alpha_{pc_i}}{H_{c_i}.|P|} \tag{12}$$

The instances are entitled with the format P*-C*-*. For example, instance P10-C10-1 defines the first instance with 10 products including 10 characteristics. Utilization Rates within each three instances with the same number of products and characteristics are different, ascending from 1 to 3. Greater UR within an instance indicates a more complex problem to solve.

Table 1. Instances' features

No.	Line	Name	# Dep const.	UR range	No.	Line	Name	# Dep const.	UR range
1	L1	P10-C10-1	2	0.30–0.75	15	L2	P18-C15-3	4	0.56–1.67
2	L1	P10-C10-2	2	0.60–1.50	16	L2	P25-C15-1	4	0.72–0.96
3	L1	P10-C10-3	2	1–1.50	17	L2	P25-C15-2	4	0.72–1.20
4	L1	P15-C10-1	2	0.53–1	18	L2	P25-C15-3	4	0.72–1.60
5	L1	P15-C10-2	2	0.50–1.07	19	L3	P10-C30-1	10	0.50–1.20
6	L1	P15-C10-3	2	0.30–1.67	20	L3	P10-C30-2	10	0.50–1.50
7	L1	P30-C10-1	2	0.20–1.20	21	L3	P10-C30-3	10	0.50–1.80
8	L1	P30-C10-2	2	0.60–1.20	22	L3	P15-C30-1	10	0.50–0.80
9	L1	P30-C10-3	2	0.80–1.20	23	L3	P15-C30-2	10	0.50–1.20
10	L2	P10-C15-1	4	0.90–1	24	L3	P15-C30-3	10	0.50–1.40
11	L2	P10-C15-2	4	1–1.20	25	L3	P20-C30-1	10	0.33–0.40
12	L2	P10-C15-3	4	1–2	26	L3	P20-C30-2	10	0.50–1.05
13	L2	P18-C15-1	4	0.44–1	27	L3	P20-C30-3	10	0.60–1.23
14	L2	P18-C15-2	4	0.56–1.11					

3.2 Computational Results

All 27 instances presented in Table 1 are computed by the three models namely CP1, CP2, and MIP. Table 2 provides the computational results including the computational time and number of violations separately sorted by dependent and independent. A run-time limitation of 3600 s is fixed for all of the computations.

Table 2. Results of solving real-world instances by models CP1, CP2, and MIP

No.	Instance	CP1	CP2			MIP [4]		
		Time (s)	*IV*	*DV*	Time (s)	*IV*	*DV*	Time (s)/gap
1	P10-C10-1	0.09 (F)	0	0	1	0	0	0.19
2	P10-C10-2	12 (INF)	3	0	19	3	0	0.13
3	P10-C10-3	10 (INF)	3	0	93	3	0	2
4	P15-C10-1	0.20 (F)	0	0	0.38	0	0	0.22
5	P15-C10-2	-	1	0	15	1	0	7
6	P15-C10-3	-	5	0	3600*	4	1	1328
7	P30-C10-1	0.28 (F)	0	0	27	0	0	0.60
8	P30-C10-2	-	0	0	141	0	0	824
9	P30-C10-3	-	1	1	3600*	2	0	100%*
10	P10-C15-1	-	0	1	48	0	1	0.44
11	P10-C15-2	-	1	1	123	1	1	0.66
12	P10-C15-3	-	6	1	2387	6	1	21
13	P18-C15-1	-	1	0	3600*	1	0	28
14	P18-C15-2	-	3	0	3600*	3	0	497
15	P18-C15-3	-	3	1	3600*	3	1	628
16	P25-C15-1	35 (F)	0	0	538	0	0	37
17	P25-C15-2	-	1	0	3600*	1	0	2401
18	P25-C15-3	-	4	0	3600*	1	0	3439
19	P10-C30-1	-	2	3	3600*	1	4	1
20	P10-C30-2	-	3	4	3600*	3	4	3
21	P10-C30-3	-	10	3	3600*	10	3	8
22	P15-C30-1	-	4	3	3600*	2	3	289
23	P15-C30-2	-	6	4	3600*	2	7	1054
24	P15-C30-3	-	8	5	3600*	5	6	1346
25	P20-C30-1	0.42 (F)	0	0	792	0	0	0.67
26	P20-C30-2	-	1	1	3600*	1	0	200
27	P20-C30-3	-	7	4	3600*	4	5	90%*

* The computational time has reached the determined limit (3600 s).
F: Feasible; INF: Infeasible; *IV*: Independent violation(s); *DV*: Dependent violation(s).

Instances in which no violations occur are feasible in CP1 and designated by "F". Conversely, instances that are infeasible in CP1 and there are no solutions without any violations are shown as "INF". The hyphen (-) in the column "Time" means that CP1 is not able to prove neither feasibility nor infeasibility within the time limit. CP1 outperforms in terms of CPU time in instances where the solution contains no violations (F). CP2 is substantially less efficient than CP1, especially in instances containing higher utilization rates or greater number of products and characteristics. Only in P30-C10-2 does CP2 perform significantly better, such that CP1 cannot even prove feasibility within the 3600-s time limit. This demonstrates that the addition of an objective function modifies the default heuristics (for variable selection, constraint propagation, etc.) of the IBM CPLEX Optimization Studio. In this particular test problem, the variables that result in an objective function of 0 after instantiation, may have been selected considerably earlier in CP2 compared to CP1.

The CP solver (with the existence of an objective function) operates in such a way that once a feasible solution is found, it can be retrieved and, if desired, persists searching for more solutions. Consequently, the search continues until all feasible solutions are checked in order to converge the best bound, or the best bound is refined after investigating various branches. In most of the experiments, CP2 considers the best bound zero, even if there are violations in the final solution. Therefore, it continues the exploration until all branches are investigated. In instance P10-C30-2, for example, a solution with the same objective function as MIP's final solution (total violations of 7) is obtained in 213 s. However, the search terminates by the time limit of 3600 s without an optimality proof. This CPU time difference between CP2 and MIP reaches its peak in instances where there are a higher number of characteristics with URs closer to the upper bounds of their ranges, as well as instances with a greater number of products and characteristics.

4 Discussion and Conclusions

In this research, we studied an extension of the Car Sequencing Problem with both ratio and Cross-Ratio constraints known as CSP-CR that takes into account the dependency between product options/characteristics as well as the independent ones. Using Constraint Programming (CP), we proposed two models of CSP-CR including: CP with hard constraints (CP1) and with soft constraints (CP2). The performance of the proposed models was compared both against each other and against the Mixed Integer Programming model (extracted from the literature) using 27 instances derived from the final assembly lines of a real-world manufacturing company.

In this study, we addressed 27 small instances, with maximum of 20 products and 30 characteristics. Among these 27 experiments, in 21 instances spacing rule violations were unavoidable. This highlights the complexity of real-world instances, even when dealing with smaller subsets of them. Furthermore, based on our extensive observations across various real-world instances on different assembly lines, where 100–150 products are produced daily, achieving a solution that strictly adheres to all spacing rules as hard constraints is unlikely in over 90% of cases. Consequently, in today's era of customized production, where spacing rule violations are almost inevitable, it is crucial to employ approaches to manage them effectively. According to our observations, there are two primary states for controlling these violations:

1. Distributing violations across different characteristics (encompassing both dependent and independent spacing rules),
2. Distribution of violations occurring on a specific characteristic (control over the congestion of violations).

In state 1, the decision maker can prioritize the occurrence of violations across the characteristics. Model CP2 accomplishes this by assigning individual weights to each spacing rule violation and minimizing them. For example, if the occurrence of violations in installing a specific characteristic lead to substantial work overloads and line stoppages, the decision maker should assign a higher weight to the violations associated with that particular characteristic.

State 2 tackles one of the limitations of State 1 (the weighting method). Our research has revealed that the way violations occur during the installation of a specific characteristic is also vital for short-term assembly line balancing. To provide further clarity, Fig. 3 depicts two different sequences of five products, three of them containing characteristic c_1. It is evident that reduced congestion of violations leads to more effective management of work overload. Hence, sequence (b) is more favorable than sequence (a). Although it contains higher quantity of violations, they are distributed more evenly. For example, it may make it easier for a worker to compensate the work overload they cause compared to sequence (a). The development of approaches to tackle this challenge is a valuable consideration for future researchers.

Slot	Product	c_1 (1:3)
1	5	1
2	3	1
3	2	
4	1	
5	4	1

(a)

Slot	Product	c_1 (1:3)
1	5	1
2	1	
3	4	1
4	2	
5	3	1

(b)

Fig. 3. Two different congestions of three work-intensive products in characteristic c_1 with independent rule (1:3); a) A sequence of products with one violation and more congestion, b) A sequence of products with two violations and less congestion

References

1. Delice, Y., Aydoğan, E.K., Himmetoğlu, S., Özcan, U.: Integrated mixed-model assembly line balancing and parts feeding with supermarkets. CIRP J. Manuf. Sci. Technol. **41**, 1–18 (2023)
2. Koren, Y., Shpitalni, M., Gu, P., Hu, S.J.: Product design for mass-individualization. Procedia Cirp **36**, 64–71 (2015)
3. Boysen, N., Fliedner, M., Scholl, A.: Sequencing mixed-model assembly lines: survey, classification and model critique. Eur. J. Oper. Res. **192**(2), 349–373 (2009)

4. Mahmoodjanloo, M., Baboli, A., Ruhla, M.: Car sequencing problem with cross-ratio constraints: a multi-start parallel local search. IFAC-PapersOnLine **55**(10), 1255–1260 (2022)
5. Brailsford, S.C., Potts, C.N., Smith, B.M.: Constraint satisfaction problems: algorithms and applications. Eur. J. Oper. Res. **119**(3), 557–581 (1999)
6. Gottlieb, J., Puchta, M., Solnon, C.: A study of greedy, local search, and ant colony optimization approaches for car sequencing problems. In: Cagnoni, S., et al. (eds.) EvoWorkshops 2003. LNCS, vol. 2611, pp. 246–257. Springer, Heidelberg (2003). https://doi.org/10.1007/3-540-36605-9_23
7. Gravel, M., Gagne, C., Price, W.L.: Review and comparison of three methods for the solution of the car sequencing problem. J. Oper. Res. Soc. **56**(11), 1287–1295 (2005)
8. Bautista, J., Pereira, J., Adenso-Díaz, B.: A beam search approach for the optimization version of the car sequencing problem. Ann. Oper. Res. **159**, 233–244 (2008)
9. Vahedi-Nouri, B., Tavakkoli-Moghaddam, R., Hanzálek, Z., Dolgui, A.: Production scheduling in a reconfigurable manufacturing system benefiting from human-robot collaboration. Int. J. Prod. Res. 1–17 (2023)
10. Butaru, M., Habbas, Z.: The car-sequencing problem as n-ary CSP–sequential and parallel solving. In: Zhang, S., Jarvis, R. (eds.) AI 2005. LNCS, vol. 3809, pp. 875–878. Springer, Heidelberg (2005). https://doi.org/10.1007/11589990_100
11. Siala, M., Hebrard, E., Huguet, M.J.: A study of constraint programming heuristics for the car-sequencing problem. Eng. Appl. Artif. Intell. **38**, 34–44 (2015)
12. Dincbas, M., Simonis, H., Van Hentenryck, P.: Solving the car-sequencing problem in constraint logic programming. In: ECAI, vol. 88, pp. 290–295 (1988)
13. Kis, T.: On the complexity of the car sequencing problem. Oper. Res. Lett. **32**(4), 331–335 (2004)
14. Bysko, S., Krystek, J., Bysko, S.: Automotive paint shop 4.0. Comput. Industr. Eng. **139**, 105546 (2020)
15. Zinflou, A., Gagné, C., Gravel, M.: Design of an Efficient Genetic Algorithm to Solve the Industrial Car Sequencing Problem. IntechOpen (2008)
16. Briant, O., Naddef, D., Mounié, G.: Greedy approach and multi-criteria simulated annealing for the car sequencing problem. Eur. J. Oper. Res. **191**(3), 993–1003 (2008)
17. Moya, I., Chica, M., Bautista, J.: Constructive metaheuristics for solving the car sequencing problem under uncertain partial demand. Comput. Ind. Eng. **137**, 106048 (2019)
18. Wu, J., Ding, Y., Shi, L.: Mathematical modeling and heuristic approaches for a multi-stage car sequencing problem. Comput. Ind. Eng. **152**, 107008 (2021)
19. Souza, F., Grimes, D., O'Sullivan, B.: Variable-relationship guided LNS for the car sequencing problem. In: Longo, L., O'Reilly, R. (eds.) AICS 2022. CCIS, vol. 1662, pp. 437–449. Springer, Cham (2023). https://doi.org/10.1007/978-3-031-26438-2_34

Picking Optimization in U-Shaped Corridors with a Movable Depot

Roberto Montemanni(✉), Agnese Cervino, and Francesco Lolli

University of Modena and Reggio Emilia, 42122 Reggio Emilia, Italy
{roberto.montemanni,francesco.lolli}@unimore.it

Abstract. We consider an order-picking system for a warehouse divided into corridors with two-layer shelves being arranged in the shape of a U in each corridor. Given an order in a corridor, the focus is on the optimization of the picking sequence and on locating the movable depot in the most convenient location. Two iterative algorithms based on constraint programming are proposed. Computational experiments position the new methods in the existing literature, showing that they are operatively effective.

We also show how allowing the depot to be allocated away from the central axis of the corridor can lead to substantial time savings, especially for small orders. This strategic option had not been considered in the previous literature, but can be easily implemented in modern warehouses.

Keywords: Warehouse management · U-shaped corridors · Picking optimization · Depot position · Constraint programming

1 Introduction

Statistics indicate that order picking accounts for 50% to 75% of the total costs to operate a warehouse [1]. Human pickers are still today carrying out picking activities in most of the scenarios, due to their still unmatchable flexibility [2,3]. However, in modern warehouses workers are supported by technology to ease their tasks, both in terms of smart vehicles to help them transporting goods, and devises with smart interfaces to drive them through the picking process. In such a case the benefit is twofold: the picker job becomes lighter both from a physical and intellectual viewpoints, while the company has an economical return associated with the enhanced efficiency. In this work we cover the use of mathematical modelling and optimization to provide pickers with efficient strategies for their picking tasks. The outcome of our approach are an order-based optimized picking sequencing and an optimized location within the corridor for the movable depot in which the picker is operating. All these information can be communicated to the workers through standard devices such as tablets.

The warehouse investigated is characterized by U-shaped corridors, where products are organized in stillages that can be stacked on top of each other for

S.-H. Sheu (Ed.): ICIEA-EU 2024, LNBIP 507, pp. 133–145, 2024.
https://doi.org/10.1007/978-3-031-58113-7_12

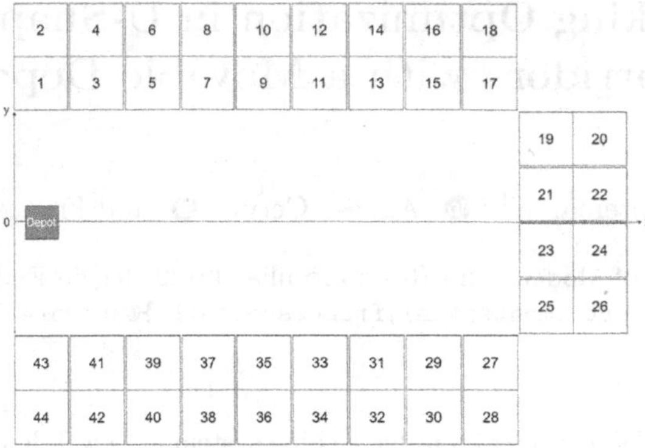

Fig. 1. Example of a U-shaped corridor with 44 stillages

a total of two layers. An example of such a corridor layout is provided in Fig. 1, where the dark grey square is the movable depot. It can be positioned based on the order characteristics before the start of the picking process.

In Sect. 2 we present a literature review on optimization in warehouses, focusing mainly on U-shaped corridors. In Sect. 3 we formally describe the problem we treat, while in Sect. 4 a constraint programming model, based on a for picking optimization is described. The optimization of the location of the depot is instead discussed in Sect. 5. Computational experiments are presented in Sect. 6, where the new methods we propose are compared with state-of-the-art algorithms from the literature. Conclusions are finally summarized in Sect. 7.

2 Literature Review

Design and operations of warehouses in the context of order picking has been vastly studied in the literature. Topics covered include layout design, storage assignment, zoning, batching and routing. General analysis and surveys on zoning, batching and routing problems can be found in [2]. The layout design problem can be aggregated at various levels. The perspective ranges from location planning [4], department arrangement inside the warehouses [5], to determining the number, orientation, and the arrangement of the shelves etc. The latter problems are highly interdependent with order picking problems, particularly in the field of storage assignment, zoning and routing. Manual picking systems are still the most popular ones in several contexts, and in [6] the average travel distance of a picker is analyzed with different routings on different layout arrangements.

A traditional layout for warehouses that consist of multiple zones with shelves organized in U-shape, is considered in [7] and in [8], where a detailed literature survey is also available. U-shaped order picking areas, as the one studied in this paper, can frequently be observed in practice, for example in the automotive or chemical industries [7]. A study on improving the performance of the U-shaped layout through pickers learning is presented in [9], while the use of semi-empty pallets that allow easier object extraction is discussed in [10]. Optimization tools are very popular to enhance efficiency in logistics ([11–13]). In [14] it is shown how to optimize layout design, storage assignment and depot location in a U-shaped corridor, both in terms of economics or ergonomics. An extension of the work, that considers more degrees of freedom for the depot position, is discussed in [15].

3 Problem Description

This paper considers order picking in a U-shaped area as outlined in [7] and [8]. In the warehouse considered items are stored in N stillages, with two rows of stillages stacked one atop the other. Each U-zone consists of two horizontal and one vertical shelf as illustrated in Fig. 1, and the geometry of the corridor can change, but is given as an input. The depot, where each order picking tour starts and ends and the collected goods are grouped, is brought to and removed from the U-zone by a forklift truck, and its position within the corridor can be decided in advance. The picker travels on foot along the shelves of the U-zone, possibly pushing or pulling a cart or a similar device.

We consider the operations of a picker while processing a single order in a single U-zone. Each order consists of a set $K \subseteq N$ of stillages that need to be visited to collect items, that have to be grouped to the depot D. For each stillage $k \in K$, a quantity d_k of goods collected is also provided (this can be weight, dimension, or a combination of the two). A capacity Q for the picking device used by the picker is also given. The picker must return to the depot when she/he has finished the order or when the transport capacity Q of the picking device has been reached. In the latter case, she/he returns to the depot to empty the picking device, and then continue the picking process. After an order has been completed, the depot is taken away. Note that although the optimization focuses on a single order, during a shift a picker processes multiple orders. We assume orders are planned beforehand, such that each order is independent from other orders in the optic of our optimization.

The U-zone's coordinate system is two-dimensional, and the depot can be placed anywhere in the U-corridor zone. The coordinates (x_D, y_D) of the depot D have to be decided before the order picking process starts, since a forklift will deposit it in the suggested position and take it away once the picking is completed. Euclidean distances are used to calculate the travel distance d_{ij} between two points of interests i with coordinates (x_i, y_i) and j with coordinates (x_j, y_j) of set $N \cup \{D\}$: $d_{ij} = \sqrt{(x_i - x_j)^2 + (y_i - y_j)^2}$, as we assume that this is the most intuitive way to travel through the pick zone. Note that only stillages containing items to be picked are considered.

When optimizing the location of the depot, the distance travelled by the forklift to position the depot is calculated from the center of the open end of the U corridor (coordinates (0,0)) to the selected location (x_D, y_D) again as a Euclidean distance, but in this case a factor ν is considered to model that this movement is carried out (twice) but the (faster) vehicle in charge of positioning and collecting the depot. The distance travelled to position and collect the depot is therefore calculated as $P_D = \nu\sqrt{(x_D)^2 + (Y_D)^2}$.

4 Sequencing of Picking Operations

Assuming the (feasible) coordinates of the depot (x_D, y_D) inside the U-corridor are given, the picking process can be optimized by solving a Capacitated Vehicle Routing Problem (CVRP) [16], as observed in [7]. Although we can not use this property in our solving method, in [8] is is proven that this problem can be further simplified, since we can consider only picking tours that visit stillages in a clockwise (or counterclockwise) order.

The picking optimization process can therefore be modelled in terms of constraint programming [17] as follows. We have a variable x_{ij} that takes value 1 if location j is visited right after location i in a picking tour, 0 otherwise. A second variable y_j is also defined for each location, containing the cumulative picking quantity collected along a tour, up to location j. It is used to model capacity constraints. Using the syntax of the CP-SAT solver of OR-tools [18], the following model can be used to describe the problem.

$$z(x_D, y_D) = P_D + \min \sum_{i \in N} \sum_{j \in N} d^D_{ij} x_{ij} \tag{1}$$

$$\textit{s.t.}\ \text{MultipleCircuit}(x_{ij} : i, j \in N) \tag{2}$$

$$x_{ij} = 1 \Rightarrow y_j = y_i + q_j \qquad i \in N, j \in K \tag{3}$$

$$x_{ij} \in \{0, 1\} \qquad i, j \in N \tag{4}$$

$$0 \leq y_i \leq Q \qquad i \in N \tag{5}$$

For notational ease, we define z as the cost of the solution of the model when the depot is in position (x_D, y_D). The objective function (1) minimizes the sum of P_D, the (scaled) distance traveled by a forklift to position the depot, and the total distance traveled by the picker. Note that distances d^D_{ij} are used to indicate that the distances are calculated with respect to the given position of the depot. Constraint (2) uses the *MultipleCircuit* statement of CP-SAT [18] to model a vehicle routing feasible solution. Constraints (3) regulate capacity. They are activated only when $x_{ij} = 1$ through the statement *OnlyEnforceIf* of CP-SAT [18], here indicated as "$\Rightarrow$", and set the cumulative weight collected up to customer j in a tour. Constraints (4) and (5) finally define the domains of the variables.

Note that the model can be solved either optimally, or heuristically by truncating the execution of the solver after a given time of T_{max} seconds. This latter option is more likely to be used for large problems, given the $\mathcal{NP}$-hard nature of the CVRP problem [16].

5 Optimization of the Depot Location

In the context of finding an optimized location for the depot, our approach consists in repeatedly solving the model constraint programming model described in Sect. 4 with different tentative locations for the depot, and choose the best. Differently from [7] and [8], where only the locations in the central axis of the corridor are considered, we decided to consider all possible positions within the corridor. The motivation for restricting the search to che center of the corridor was related to the maneuvering space for the picking devices assisting the workers. In our experience, however, warehouses are nowadays equipped with agile picking devices able to move around easily, so we decided to open for more options for the depot location. Note that in case the optimization has to be restricted to the central axis of the corridor, as in [7] and [8], it is enough to set the feasible coordinates to 0 only for the y axis (see Fig. 1).

We will discuss two algorithms, the first regularly scanning all the area with a given step, and the second focussing the search on the promising locations only, through a zooming mechanism.

5.1 The Basic Algorithm

In this method, given a precision step s, all the locations on a grid defined over the feasible area for the depot, and with a step s, will be examined. A visual example is provided in Fig. 2, where the red points are examined.

The pseudocode of the *Basic* approach is provided in Algorithm 1. The method takes in input the coordinates of the top-left and bottom-right corner of the (feasible) search area for the depot location together with the search-step s. After an initialization phase (lines 1–3), the main double loop is entered and for each point of the grid defined by s, the model described in Sect. 4 is solved with the given coordinates for the depot for a maximum of T_{max} seconds. In case the solution is better than the best one retrieved so far, the data of the best solution are updated (line 7 and 8). The method returns the coordinates of the best location for the depot and the relative cost.

5.2 The ZoomIn Algorithm

A major drawback of the *Basic Algorithm* discussed in Sect. 5.1 is that the exploration for the best depot location is carried out evenly along the whole search area. In the reality, nearby locations will tend to give similar costs, and the costs will gradually smoother into different values as the distance increases. Such a situation is depicted in Fig. 3 for the instance *44-(9x4)-10-01*, where *1D* stands for the locations along the central axis ($y = 0$) only and *2D* for all locations; green areas indicate locations with a lower cost and the red ones locations with a higher cost. We can exploit this situation and zooming only in promising areas, aiming at reducing the overall computation time of the search.

The search area for the depot location is first analyzed with a large step s_M and only the surrounding of the most promising location is selected for a

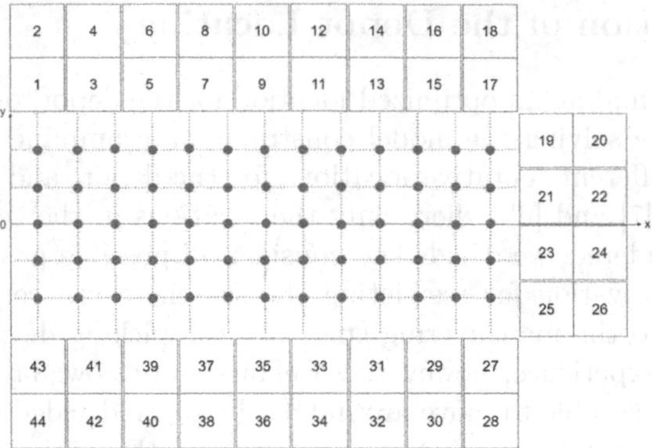

Fig. 2. Depot positions scanned by the *Basic* algorithm with a given step

Algorithm 1 - Basic($(x_m, y_m), (x_M, y_M), s$)

```
(x^B, y^B) = (x_m, y_m)
z^B = +∞
for y_D in range(y_m, y_M, s) do
    for x_D in range(x_m, x_M, s) do
        Solve the model from Section 4 for T_max seconds
        if z(x_D, y_D) < z^B then
            (x^B, y^B) = (x_D, y_D)
            z^B = z(x_D, y_D)
        end if
    end for
end for
return (x^B, y^B), z^B
```

deeper analysis, carried out with a smaller step. This process is continued until the required precision is reached. It is worth however observing that such a procedure makes the search process more heurristic. Looking again to Fig. 3, for example, a crude initial analysis might lead to search in details the top-right green area, which unfortunately can only lead to a local optimum.

The pseudocode for the algorithm *ZoomIn* is provided in Algorithm 2. the method takes in input the coordinates of the top-left corner and bottom-right corners of the search area for the depot location, together with the initial step value s_M, the final step value s_m and the reduction factor e_s for the step.

The algorithm starts by setting the working boundaries (lines 1 and 2) and the working value of s (line 3). Then the main loop is entered where the *Basic* algorithm is invoked with the working parameters (line 5). The working boundaries are set to the surrounding of the best solution retrieved (lines 6 and 7). Note the use of the min and max operators to constraint the optimization to remain within the feasible region for the depot. The value of s is finally reduced at line 8, to augment the precision in the next iteration. Note that the main

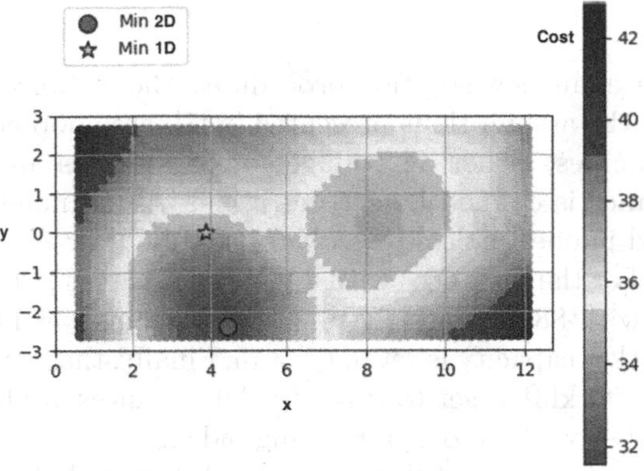

Fig. 3. Heatmap for the distance travelled by the picker for different locations of the depot. Instance *44-(9x4)-10-01*, s = 0.2, t=10 s

Algorithm 2 - ZoomIn$((x_m, y_m), (x_M, y_M), s_M, s_m, e_s)$

$(\overline{x}_m, \overline{y}_m) = (x_m, y_m)$
$(\overline{x}_M, \overline{y}_M) = (x_M, y_M)$
$s = s_M$
while $s \geq s_m$ **do**
    (x^B, y^B), z^B=Basic$((\overline{x}_m, \overline{x}_M), (\overline{y}_m, \overline{y}_M), s)$
    $(\overline{x}_m, \overline{y}_m) = (\max\{x_m, x^B - s\}, \max\{y_m, y^B - s\})$
    $(\overline{x}_M, \overline{y}_M) = (\min\{x_M, x^B + s\}, \min\{y_M, y^B + s\})$
    $s = s/e_s$
end while
return (x^B, y^B), z^B

while loop is exited once the desired precision s_m has been reached. The algorithm finally returns the best coordinates retrieved for the location of the depot, and the relative cost.

6 Computational Experiments

The instances used for the experiments are first introduced in Sect. 6.1, The methods we propose are then compared with the existing ones while allowing the depot to be positioned only on the central axis of the corridor ($y = 0$ – *1D* in the remainder of the paper). The analysis is then extended to the case where the depot can be located everywhere inside the corridor (*2D* in the remainder of the paper), aiming at understanding in which measure the extra freedom can lead to more optimized solutions.

6.1 Instances

In order to evaluate our new solution procedures, the instances proposed in [8], that are based on the assumptions presented in [7], are adopted. U-layouts with two different capacities, either 44 stillages or 88 stillages are considered. The layout of an instance is defined by setting n and m, the number of stillages in one horizontal and in one vertical row. We randomly draw K items to be picked for each instance. For the instances with 44 stillages, we set either $|K| \in \{10, 15\}$, for the instances with 88 stillages $|K| \in \{30, 60\}$. Items weights $q_j \in \{1, \ldots, 5\}$ $\forall j \in K$ and the picker capacity is set to $Q = 15$. Finally, the factor for positioning the depot with the forklift is set to $\nu = 1/3$. All instances are labeled as follows: $|K| - (n, m) - |K| - \alpha$ where α is a running index.

The measurements of the stillages to $w = 1.3$ m and the gap between the stillages to $s = 0.05$ m. A minimum distance of 0.65 m has to exist between the depot location and the stillages (or the entrance of the corridor), in order to allow enough working space around the depot itself. The coordinates of a stillage (or the depot) used for the calculation of the distances are given by its center.

The reader interested in a complete description of the derivation of the instances can refer to [8], The instances themselves are available for download at https://doi.org/10.5281/zenodo.4671870.

6.2 1D Case: Comparison with State-of-the-Art Algorithms

In this section we consider the case 1D, where the depot is constrained to be over the central axis of the corridor ($y = 0$), as previously considered in the literature. The results of the algorithms described in Sect. 5 are compared with those of the methods previously appeared in the literature. The following methods are considered:

- *Sweep* from [7]. It was implemented in Mathematica 7.0 [19]. No information about the computer used for the experiments was provided.
- *Benders Decomp.* from [8]. It was coded in C#. The experiments were run on a computer equipped with 8 GB of RAM and an Intel Core i7-3631QM CPU, with CPLEX 12.10 [20] used to solve linear programs. A maximum running time of 3600 s is allowed (a dash in the column *Sec* means that this limit is reached).
- *Dynamic Progr.* from [8]. It was coded in C#. The experiments were run on a computer equipped with 8 GB of RAM and an Intel Core i7-3631QM CPU.
- *Basic* (Sect. 5.1) and *ZoomIn*. It was coded in Python 3. The experiments were run on a computer equipped with 32 GB of RAM and an Intel Core i7 12700F CPU, with OR-Tools CP-SAT 9.6 [18] used as a solver for Constraint Programming model.
- *ZoomIn* (Sect. 5.2). It was coded in Python 3. The experiments were run on a computer equipped with 32 GB of RAM and an Intel Core i7 12700F CPU, with OR-Tools CP-SAT 9.6 [18] used as a solver for Constraint Programming model.

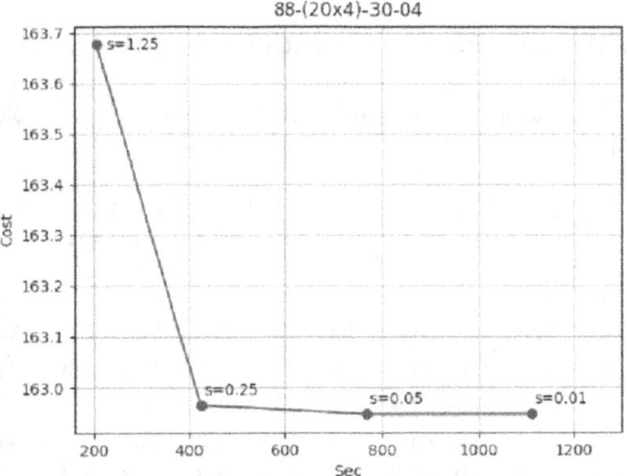

Fig. 4. Algorithm *ZoomIn* with $s_M = 1.25$, $s_m = 0.01$, $e_s = 5$ and $T_{max} = 3$: best cost over different values of s for instance *88-(20x4)-30-04*

The results are summarized in Table 1, where for each instance we report the best cost obtained by the method together with the computation time required. Note that for the algorithms we propose, experiments with different parameter settings are proposed (they are indicated in the table). This suggests that it is possible to tune these methods to me either faster or more precise.

From Table 1 it emerges that the new constraint programming-based methods we propose are competitive when compared with the existing algorithms. Although they are slower than other approaches, they provide high quality results. In particular the *ZoomIn* method match or improves all the previous best-known results, keeping however the computation times below 550 s in the worst case. Such a time is practically sufficient for a Company to plan in advance the pickings for the next day, assuming to have proper hardware, like a suitable multicore architecture.

When comparing the *Basic* and *ZoomIn* methods, the convenience of the latter emerges clearly, since a much smaller (final) step can be achieved in comparable computation times. However, the it can be observed that the gain in the total distance does not improve much when the step becomes very small. This is evident in Fig. 4 where the evolution of the total distance over time is reported. This suggest that settings where the final step is not too small should be preferred to have a better trade off between computation time and quality of the solution. This intuition will be followed for the results reported in Sect. 6.3.

6.3 2D Case: Allowing the Depot Out of the Central Axis of the Corridor

In this section we compare the best results obtained in Sect. 6.2 while considering only locations for the depot along the central axis, with new results obtained allowing the depot to be placed anywhere in the corridor. Our aim is to

Table 1. 1D case: comparison with state-of-the-art algorithms

Instance	Benders Decomp. [8]		Sweep [7]		Dynamic Progr. [8]		Basic Algorithm				ZoomIn Algorithm			
							$s = 1$ $T_{max} = 3$		$s = 1$ $T_{max} = 10$		$s_M = 1.25$ $s_m = 0.01$ $e_s = 5$ $T_{max} = 3$		$s_M = 1.25$ $s_m = 0.01$ $e_s = 5$ $T_{max} = 10$	
	Val	Sec	Val	Sec	Val	Sec	Val	Sec	Val	Sec	Val	Sec	Val	Sec
44-(9 × 4)-10-01	34.3	0.6	34.3	0.0	34.3	0.8	34.4	1.4	34.4	1.3	34.3	3.9	34.3	3.9
44-(9 × 4)-10-02	37.4	0.3	37.4	0.0	37.4	0.3	37.5	3.6	37.5	3.7	37.4	5.7	37.4	5.3
44-(9 × 4)-10-03	44.6	1.4	45.5	0.0	44.6	0.2	44.6	3.3	44.6	3.3	44.6	4.9	44.6	4.5
44-(9 × 4)-10-04	37.3	0.5	37.3	0.0	37.3	0.2	37.3	3.3	37.3	3.2	37.3	5.8	37.3	5.9
44-(9 × 4)-10-05	32.7	0.3	32.7	0.0	32.7	0.7	32.7	3.2	32.7	3.5	32.7	6.0	32.7	5.9
44-(9 × 4)-10-06	39.0	0.7	39.0	0.0	39.0	0.6	39.1	0.9	39.1	1.0	39.0	2.7	39.0	2.4
44-(9 × 4)-10-07	43.1	1.1	44.0	0.0	43.1	0.4	43.1	2.1	43.1	2.1	43.1	5.4	43.1	5.5
44-(9 × 4)-10-08	37.5	0.6	37.9	0.0	37.5	0.4	37.5	7.1	37.5	6.9	37.5	10.3	37.5	9.7
44-(9 × 4)-10-09	34.4	0.4	35.6	0.0	34.4	0.3	34.4	5.3	34.4	5.5	34.4	6.0	34.4	6.5
44-(9 × 4)-10-10	34.4	0.7	34.4	0.0	34.4	0.2	34.5	9.0	34.5	10.2	34.4	15.8	34.4	19.5
44-(9 × 4)-15-01	52.4	39.3	52.4	0.0	52.4	0.6	52.4	36.5	52.4	116.8	52.4	124.6	52.4	407.8
44-(9 × 4)-15-02	57.4	54.8	57.4	0.0	57.4	0.4	57.6	36.4	57.6	120.5	57.4	121.5	57.4	401.7
44-(9 × 4)-15-03	53.8	99.5	56.4	0.0	56.4	0.6	53.8	36.5	53.8	120.5	53.8	121.5	53.8	401.7
44-(9 × 4)-15-04	47.4	12.7	47.4	0.0	47.4	0.9	47.4	36.5	47.4	102.8	47.4	121.5	47.4	219.3
44-(9 × 4)-15-05	42.5	8.1	42.5	0.0	42.5	1.4	42.5	31.1	42.5	63.5	42.5	78.3	42.5	105.7
44-(9 × 4)-15-06	49.4	19.1	49.4	0.0	49.4	0.4	49.4	36.4	49.4	120.5	49.4	121.6	49.4	401.6
44-(9 × 4)-15-07	43.7	13.1	43.7	0.0	43.7	1.0	43.7	36.5	43.7	119.8	43.7	121.5	43.7	398.2
44-(9 × 4)-15-08	47.0	7.5	47.0	0.0	47.0	0.6	47.0	36.5	47.0	94.0	47.0	121.6	47.0	315.6
44-(9 × 4)-15-09	53.0	79.9	53.0	0.0	53.0	0.4	53.1	36.4	53.1	120.5	53.0	121.6	53.0	401.6
44-(9 × 4)-15-10	45.8	16.8	46.9	0.0	45.8	1.0	45.8	33.9	45.8	91.0	45.8	73.5	45.8	123.4
88-(20 × 4)-30-01	148.9	–	145.7	0.2	144.0	4.7	142.0	83.4	141.4	272.7	141.4	163.7	141.3	535.1
88-(20 × 4)-30-02	148.0	–	140.9	0.2	140.9	2.9	141.7	83.5	139.2	272.8	139.2	164.1	139.1	535.4
88-(20 × 4)-30-03	146.7	–	139.3	0.2	139.3	3.5	132.6	83.2	131.9	272.6	131.8	163.5	131.8	535.2
88-(20 × 4)-30-04	177.2	–	167.8	0.2	166.4	2.2	164.1	83.5	163.0	272.7	163.0	163.8	163.0	535.2
88-(20 × 4)-30-05	145.5	–	142.0	0.2	138.1	2.5	136.8	83.5	136.6	272.8	136.6	163.7	136.5	535.5
88-(20 × 4)-30-06	172.0	–	159.8	0.2	159.8	2.7	155.1	83.5	155.1	272.9	154.9	164.0	154.9	535.5
88-(20 × 4)-30-07	143.1	–	134.5	0.2	134.5	6.7	135.4	83.3	132.5	272.6	132.7	163.5	132.5	535.4
88-(20 × 4)-30-08	151.9	–	147.2	0.2	147.2	4.0	148.1	83.6	147.2	272.8	147.2	163.8	147.2	535.1
88-(20 × 4)-30-09	154.8	–	148.0	0.2	148.0	2.8	146.9	83.4	146.5	272.6	146.5	163.8	146.4	535.2
88-(20 × 4)-30-10	144.0	–	137.9	0.2	137.9	3.5	138.0	83.3	138.0	272.8	137.9	163.9	137.9	535.2
88-(20 × 4)-60-01	297.2	–	255.6	0.7	255.6	6.2	280.0	87.4	255.7	277.2	263.5	171.2	254.3	543.5
88-(20 × 4)-60-02	305.8	–	237.2	0.7	234.6	8.3	249.5	88.1	239.5	277.6	256.8	173.0	229.3	544.1
88-(20 × 4)-60-03	288.5	–	235.0	0.7	231.6	10.4	245.1	87.8	230.1	277.9	239.0	172.4	228.5	544.0
88-(20 × 4)-60-04	376.9	–	285.3	0.8	283.1	6.3	304.8	87.9	275.0	278.0	286.0	172.1	270.3	544.5
88-(20 × 4)-60-05	309.4	–	239.1	0.7	238.0	6.2	245.8	87.6	236.8	277.5	237.6	171.5	235.7	545.2
88-(20 × 4)-60-06	341.6	–	266.1	0.8	263.7	5.8	280.7	87.6	261.5	277.0	277.1	172.0	258.0	544.5
88-(20 × 4)-60-07	339.6	–	270.4	0.8	268.3	5.1	290.1	87.7	267.5	277.4	271.0	171.6	265.8	544.0
88-(20 × 4)-60-08	334.9	–	275.3	0.8	274.1	5.2	297.4	87.5	263.4	277.7	279.8	171.8	257.4	544.4
88-(20 × 4)-60-09	310.4	–	240.3	0.8	239.9	6.0	250.4	87.9	241.6	277.5	240.5	172.5	235.9	544.5
88-(20 × 4)-60-10	293.2	–	252.9	0.7	251.7	9.4	270.7	87.7	252.5	277.8	261.5	172.4	250.0	544.5

Table 2. 2D case: allowing the depot out of the central axis of the corridor

Instance	1D Best	2D ZoomIn $s_M = 2$, $s_m = 0.5$ $e_s = 5$, $T_{max} = 20$			Instance	1D Best	2D ZoomIn $s_M = 2$, $s_m = 0.5$ $e_s = 5$, $T_{max} = 20$		
	Val	Val	Sec	% Imp		Val	Val	Sec	% Imp
44-(9 × 4)-10-01	34.3	31.3	2.7	8.7	88-(20 × 4)-30-01	141.3	141.3	542.2	0.0
44-(9 × 4)-10-02	37.4	37.3	4.8	0.3	88-(20 × 4)-30-02	139.1	136.6	542.4	1.8
44-(9 × 4)-10-03	44.6	42.7	3.0	4.2	88-(20 × 4)-30-03	131.8	131.8	542.4	0.0
44-(9 × 4)-10-04	37.3	36.6	3.7	1.8	88-(20 × 4)-30-04	163.0	163.0	542.5	0.0
44-(9 × 4)-10-05	32.7	29.4	1.4	10.2	88-(20 × 4)-30-05	136.5	135.6	542.5	0.7
44-(9 × 4)-10-06	39.0	37.5	1.4	3.9	88-(20 × 4)-30-06	154.9	152.8	542.4	1.4
44-(9 × 4)-10-07	43.1	42.1	2.6	2.3	88-(20 × 4)-30-07	132.5	128.9	542.4	2.7
44-(9 × 4)-10-08	37.5	35.5	7.3	5.3	88-(20 × 4)-30-08	147.2	146.6	542.4	0.4
44-(9 × 4)-10-09	34.4	34.4	5.1	0.0	88-(20 × 4)-30-09	146.4	145.6	542.6	0.6
44-(9 × 4)-10-10	34.4	31.3	10.1	8.8	88-(20 × 4)-30-10	137.9	137.8	542.6	0.1
44-(9 × 4)-15-01	52.4	51.9	366.6	0.9	88-(20 × 4)-60-01	254.3	249.2	547.6	2.0
44-(9 × 4)-15-02	57.4	57.3	363.5	0.2	88-(20 × 4)-60-02	229.3	229.4	547.7	0.0
44-(9 × 4)-15-03	53.8	53.1	425.0	1.2	88-(20 × 4)-60-03	228.5	225.3	547.7	1.4
44-(9 × 4)-15-04	47.4	45.4	117.4	4.3	88-(20 × 4)-60-04	270.3	265.1	548.0	1.9
44-(9 × 4)-15-05	42.5	41.5	46.5	2.2	88-(20 × 4)-60-05	235.7	232.6	547.8	1.3
44-(9 × 4)-15-06	49.4	47.0	408.0	4.8	88-(20 × 4)-60-06	258.0	258.0	548.0	0.0
44-(9 × 4)-15-07	43.7	43.1	222.1	1.2	88-(20 × 4)-60-07	265.8	264.7	548.0	0.4
44-(9 × 4)-15-08	47.0	46.9	190.5	0.1	88-(20 × 4)-60-08	257.4	253.0	548.0	1.7
44-(9 × 4)-15-09	53.0	50.6	364.2	4.6	88-(20 × 4)-60-09	235.9	235.6	547.6	0.1
44-(9 × 4)-15-10	45.8	45.7	184.2	0.1	88-(20 × 4)-60-10	250.0	247.0	547.5	1.2

understand how much this extra degree of freedom can enhance the quality of the solutions.

The results are summarized in Table 2, where the best known results for the 1D case (from Table 1) are shown together with the results achieved by the 2D-version of the *ZoomIn* algorithm with settings allowing to conclude the computation within 550 s for each instance. A column reporting the improvement of the 2D solution with respect to the 1D one is also incorporated in the table.

The results of Table 2 suggest that considering the possible location of the depot in the whole corridor (2D) and not only in the center (1D) can lead to substantial improvements on the total distance travelled, especially for those orders that require picking from a small proportion of the total stillages. For example, instance *4-(9x4)-10-05* where only 10 stillages are visited over the 44 of the corridor, moving the depot on one side leads to a shortening of the picking process of 10.2%. The advantage is less evident for orders that have to touch most of the stillages (like the last instances, in which 60 stillages over 88 have to be visited).

An average advantage of 2.1% has been obtained over the given instances by moving the depot out of the central axis of the corridor. Given that computation times are in the order of those of the 1D case, we conclude that positioning the depot off-center should be considered by Companies to increase their efficiency.

7 Conclusion

We considered a warehouse with U-shaped corridors and we treated the optimization of the single-order picking process. In particular, the location of the depot is considered as part of the optimization.

From an operational viewpoint, the constraint programming-based algorithms we propose are able to improve the results of the methods available from the literature, although we longer (but still feasible) computation times.

From a strategical perspective, we also show that positioning the depot off-centered with respect to the central axis of the corridor – as it was common in the previous literature – can enhance the overall picking time. This phenomenon is particularly clear for instances where picking involves only a relative small fraction of the stillages. A further study positioning this result from a more strategic point of view is foreseen.

References

1. Coyle, J.J., Bardi, E.J., Langley, C.J.: The Management of Business Logistics. West Publishing Company, St Paul, MN (1996)
2. De Koster, R., Le-Duc, T., Roodbergen, K.J.: Design and control of warehouse order picking: a literature review. Eur. J. Oper. Res. **182**(2), 481–501 (2007)
3. Ognibene Pietri, N., Chou, X., Loske, D., Klumpp, M., Montemanni, R.: The buy-online-pick-up-in-store retailing model: optimization strategies for in-store picking and packing. Algorithms **14**(12), 350 (2021)
4. Kusiak, A., Heragu, S.S.: The facility layout problem. Eur. J. Oper. Res. **29**(3), 229–251 (1987)
5. Heragu, S.S., Du, L., Mantel, R.J., Schuur, P.C.: Mathematical model for warehouse design and product allocation. Int. J. Prod. Res. **43**(2), 327–338 (2005)
6. Roodbergen, K.J., Sharp, G.P., Vis, I.F.: Designing the layout structure of manual order picking areas in warehouses. IIE Trans. **40**(11), 1032–1045 (2008)
7. Glock, C.H., Grosse, E.H.: Storage policies and order picking strategies in U-shaped order-picking systems with a movable base. Int. J. Prod. Res. **50**(16), 4344–4357 (2012)
8. Diefenbach, H., Emde, S., Glock, C.H., Grosse, E.H.: New solution procedures for the order picker routing problem in u-shaped pick areas with a movable depot. OR Spectr. **44**, 535–573 (2022)
9. Grosse, E.H., Glock, C.H.: An experimental investigation of learning effects in order picking systems. J. Manuf. Technol. Manag. **24**(6), 850–872 (2013)

10. Glock, C.H., Grosse, E.H., Abedinnia, H., Emde, S.: An integrated model to improve ergonomic and economic performance in order picking by rotating pallets. Eur. J. Oper. Res. **273**(2), 516–534 (2019)
11. Chou, X., Gambardella, L., Montemanni, R.: A Tabu search algorithm for the probabilistic orienteering problem. Comput. Oper. Res. **126**, 105107 (2021)
12. Dell'Amico, M., Montemanni, R., Novellani, S.: Algorithms based on branch and bound for the flying sidekick traveling salesman problem. Omega **104**, 102493 (2021)
13. Montemanni, R., Smith, D., Gambardella, L.: Ant colony systems for large sequential ordering problems. In: Proceedings of the IEEE Symposium on Swarm Intelligence (SIS), pp. 60–67 (2007)
14. Diefenbach, H., Glock, C.H.: Ergonomic and economic optimization of layout and item assignment of a u-shaped order picking zone. Comput. Ind. Eng. **138**, 106094 (2019)
15. Montemanni, R., Landolfo, A., Chou, X., Loske, D., Klumpp, M.: Ergonomics and storage base position for u-shaped picking zones. In: Proceedings of the 9th International Conference on Industrial Engineering and Applications (ICIEAeu), pp. 199–206. ACM (2023)
16. Toklu, N., Montemanni, R., Gambardella, L.: An ant colony system for the capacitated vehicle routing problem with uncertain travel costs. In: Proceedings of the IEEE Symposium on Swarm Intelligence (SIS), pp. 32–39 (2013)
17. Montemanni, R., Dell'Amico, M.: Solving the parallel drone scheduling traveling salesman problem via constraint programming. Algorithms **16**(1), 40 (2023)
18. Google: OR-Tools. https://developers.google.com/optimization/
19. Wolfram: Mathematica (2023). https://www.wolfram.com/mathematica/
20. ILOG, I.: User's manual for CPLEX (2023). https://www.cplex.com/

Sustainable and Energy-Efficient Industrial Systems: Modelling the Environmental Impact of Logistics Facilities

S. Perotti[1(✉)], L. Cannava[1], B. Najafi[2], E. Gronda[1,2], and F. Rinaldi[2]

[1] Department of Management, Economics and Industrial Engineering, Politecnico di Milano, 20156 Milan, MI, Italy
{sara.perotti,luca.cannava}@polimi.it, edoardo.gronda@mail.polimi.it

[2] Department of Energy, Politecnico di Milano, 20156 Milan, MI, Italy
{behzad.najafi,fabio.rinaldi}@polimi.it

Abstract. In the last decade industrial systems have been affected by increased challenges, and among those the need for more sustainable production and logistics has been called into question by both practitioners and academia. Within industrial networks, logistics facilities traditionally represent key nodes as they have a direct impact on both companies' service levels and logistics costs. Recently, their complexity has dramatically increased due to their ever-demanding requirements and pressures from stakeholders and society. In this context, companies have started to search for solutions for greener warehousing processes and energy efficiency improvements. Still, on the academic side, a limited number of studies have been found addressing the quantification of logistics facilities' environmental performance, the impact of the green warehousing practices in place, and the related effects on warehouse consumption and emission reduction. This paper aims to address this research gap by proposing a simulation-based approach where multiple scenarios of a real logistics facility are discussed, grounded on a conceptual framework that offers a roadmap towards sustainable and energy-efficient warehousing. Different scenarios are outlined, and the related performances are examined in terms of energy consumption and $CO2_{eq}$ emissions. Implications of the results are discussed and streams for future investigation are identified.

Keywords: Environmental sustainability · Sustainable warehousing · Energy efficiency · GHG emissions · Simulation

1 Introduction

Warehouses play a critical role in maintaining service levels, managing costs, and serving as central nodes within industrial systems for storing, sorting, and distributing goods [1]. In the last decade, industrial systems have been affected by increased challenges, and among those the need for more sustainable production and logistics has received increasing attention. Regarding the logistics operations, traditionally, efforts to reduce

S.-H. Sheu (Ed.): ICIEA-EU 2024, LNBIP 507, pp. 146–157, 2024.
https://doi.org/10.1007/978-3-031-58113-7_13

emissions have primarily been focused on transport activities due to their significant contribution to the companies' overall environmental impact [2]. However, a growing awareness has started to emerge on the role of logistics facilities due to their energy-intensive nature ([3, 4]). This has induced companies to look at the implementation of energy efficiency measures and green warehousing solutions [5]. In this context, Bartolini et al. [6] defined green warehousing (GW) as "a managerial concept integrating environmentally conscious solutions to minimize energy consumption, reduce energy costs, and mitigate emissions within warehouse environments". This concept is applied to any logistics building regardless of the corresponding market, operating conditions, and structural of structure. Nevertheless, conditioned logistics warehouses have emerged as crucial components of industrial networks in terms of their environmental impact, due to the large amount of energy that is required to uphold multiple temperature zones within a single warehouse environment [7]. Thus, conditioned warehouses are exceptionally complex to operate in terms of energy efficiency and environmental impact, due to their demanding logistics requirements and limitations [8].

Consequently, the implementation of GW practices in conditioned warehouses are critical to achieve Net Zero Emission goals. Moreover, literature still lacks empirical studies to assess the impact of GW practices on environmental impact and energy consumption in logistics facilities, especially in conditioned warehouses [4].

To fill this gap, this study proposes a simulation-based approach, supported by empirical data, to examine the consequences of GW practice implementation in a conditioned logistics facility. More specifically, it explores how these practices impact energy consumption by conducting multiple simulation scenarios using a real conditioned logistics facility located in Northern Italy. The simulation-based approach is underpinned by a conceptual framework that highlight the GW practices selection process. Starting from an initial analysis of the base case, each alternative scenario is systematically explored through an iterative process to identify and assess the most suitable GW measures in terms of energy efficiency and environmental performance.

Accordingly, the paper is structured as follows: The next section provides a review of the existing literature related to GW practices investigated in the simulated scenarios. Subsequently, the proposed conceptual framework and employed methodology are the described, followed by an analysis on the results of each investigated scenario. Finally, conclusions are addressed, limitations of the study are highlighted, and recommendations for future research are provided.

2 Research Background

2.1 Lighting Systems in Logistics Facilities

Lighting has traditionally been identified as a significant contributor to energy consumption in warehouses. However, the literature lacks a consistent estimate of its exact share [9]. Some scholars posit that lighting alone may account for as much as 65% of the total energy consumption in ambient-temperature (non-conditioned) warehouses [10], while this share drops to a value of 5–10% in the case of conditioned warehouses. LED lighting holds remarkable promise for energy savings and has already gained popularity as the preferred lighting solution within the warehousing sector [11]. LED lighting adoption not only offers substantial energy savings but also aligns with the contemporary trend toward more adaptable and environmentally conscious lighting solutions.

2.2 Warehouse Heating and Cooling Through Heat Pumps

Heat pumps can operate for both heating and cooling During the winter season, they can extract heat from the outdoor air, ground, or water sources and transfer it into the building. On the other hand, they can extract heat from the indoor air and transfer it outside during summer season, effectively cooling the building. Heat pumps are considered a promising technology for conditioned logistics facilities due to their high efficiencies compared to fossil fuel combustion, resulting in energy savings and reduced greenhouse gas emissions ([12, 13]). Moreover, an additional benefit of the heat pump lies in the electrification of the heating technology, for which gas-fed systems are conventionally used.

2.3 Building Envelope's Thermal Insulation

Thermal insulation is a crucial aspect of building design, particularly in conditioned logistics facilities maintaining a stable internal temperature (i.e., by reducing heat exchange with the external environment) is essential [14]. According to Paraschiv et al. [15], thermal insulation on building walls can lead to a reduction in energy consumption of approximately 13–16% depending on variations in outside air temperature, thus contributing to decrease the resulting GHG emissions of the HVAC (heating, ventilation and air conditioning) system.

3 Methodology

Several companies have committed to achieving net-zero emissions goals. To support practitioners throughout their decarbonisation path and address an identified lack in the extant literature, a framework has been outlined as per Fig. 1. The framework is intended to provide guidance in the selection of the most suitable GW practices given the specific requirements and existing constraints of the industrial system and logistics facility under examination, and to assess their impact from an energy, environmental and cost-effectiveness perspective. The framework was developed based on both academic literature and secondary sources, and further refined by means of empirical analysis. The research approach combines deductive and inductive methods, in line with Siems et al. [16], allowing flexibility for application across logistics facilities varying industries, sizes, and operating conditions, without reliance on specific simulation software.

The identification process of GW practices requires a closed-loop simulation process with a priority-based approach. Each practice is selected based on its impact on end-use types, related environmental (i.e., $CO2_{eq}$) and economic KPIs in order to implement the most convenient measures in terms of energy efficiency improvement and environmental impact, without neglecting the related economic benefits. Design Builder has been selected as the modelling and simulation software, coherently with previous studies in the literature that were dedicated to building energy simulation in other sectors [17]. Design Builder (based on Energy Plus engine) is a comprehensive software for energy

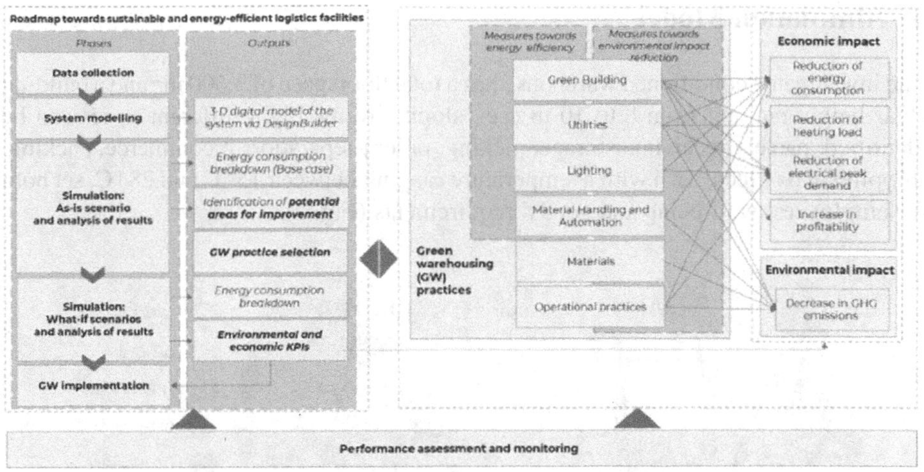

Fig. 1. Sustainable and energy-efficient industrial systems: framework for GW practice selection and implementation at logistics facilities. Adapted from: Perotti and Colicchia (2023).

simulations and energy analysis. It encompasses detailed 3D digital model allowing the assessment of energy consumption and $CO2_{eq}$ emissions through simulating alternative scenarios.

For the application of the proposed framework, a conditioned logistics facility of a cosmetics distributor located in Northern Italy was selected. In accordance with the proposed methodological framework, the 3D model of the logistics facility was developed leveraging data concerning the corresponding employed units, operational factors, and structural features of the building, gathered directly from semi-structured interviews and ad hoc on-site visits.

A base case scenario was first simulated, including all the GW practices already implemented by the company (i.e., photovoltaic (PV) panels on rooftop, LED lighting system in some areas of the logistics facility). Starting from the base case, four alternative scenarios were then simulated:

1. Scenario A: Installation of LED bulbs throughout the entire logistics facility,
2. Scenario B: Heat pump system's implementation,
3. Scenario C: Thermal insulation of the logistics facility (i.e., building envelope),
4. Scenario D: Building envelope's thermal insulation and heat pump system implementation.

For each scenario the resulting yearly energy consumption, disaggregated the corresponding energy sources (i.e., electricity and natural gas), the average monthly self-consumption ratio, the corresponding $CO2_{eq}$ emissions, and economic KPIs related to the implemented GW practices were then presented and discussed.

4 Simulation Model

The investigated conditioned warehouse has a total floorspace of 3,500 m^2 and a building clear height ranging from 8 to 10 m (i.e., sloping roof). Eight different zones can be identified, namely receiving, storage, picking, order preparation, e-commerce, packing, shipping and offices, each with a temperature ranging between 15 °C and 28 °C, set both for employees' well-being and goods' requirements (Fig. 2).

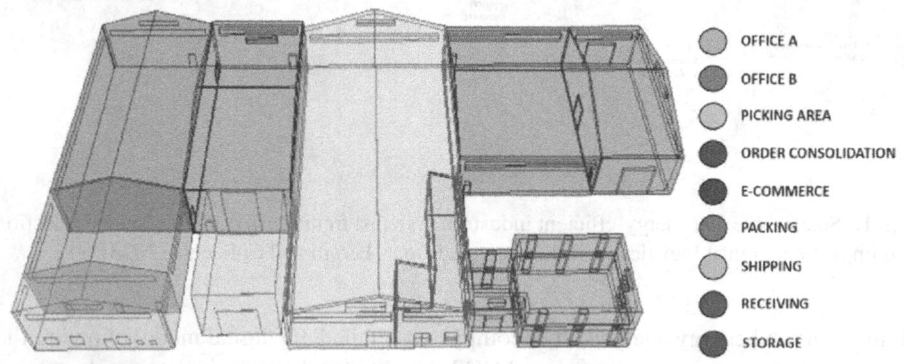

Fig. 2. Warehouse representation by zone

4.1 Base Case Scenario

Artificial lighting is provided by LED and fluorescent bulbs in line with the illuminance levels (lux) identified by the UNI EN 12464–1 standard. According to Füchtenhans et al. [10], LED bulbs luminous efficiency is estimated to be approximately equal to 100 lm/W in warehouse areas for residential use (e.g., offices) and 200 lm/W in industrial ones (i.e., all the others). On the other hand, the luminous efficiency of fluorescent bulbs is estimated to be approximately equal to 50 lm/W and 100 lm/W for offices and industrial zones respectively (Table 1).

Table 1. Lighting requirements for each warehouse zone

Zone	Floorspace [m^2]	Illuminance [lux]	Lighting power [kW]
Office A	345.0	500	1.73
Office B	305.7	500	3.06
Picking	917.2	200	0.92
Order consolidation	414.6	200	0.41

(continued)

Table 1. (*continued*)

Zone	Floorspace [m^2]	Illuminance [lux]	Lighting power [kW]
E-Commerce	509.2	200	0.51
Packing	197.8	300	0.30
Shipping	138.4	200	0.14
Receiving	259.1	200	0.52
Storage	458.9	150	0.69

Material handling is executed manually by using forklift trucks powered by lead-acid batteries. Table 2 provides a detailed overview on size and features of the material handling equipment (MHE) fleet, as well as energy consumption required by charging operations.

Table 2. Size and features of MHE fleet

Type	Fleet size [No.]	Battery capacity [kWh]	Avg. Charging time [h]	Avg. Charging rate [kWh/h]	Charging Frequency [charges/day]	Total energy consumption [kWh/day]
Reach Trucks	2	20.8	8	2.6	1	41.6
Counterbalanced	2	29.1	8	3.6	1	58.2

To perform the above-mentioned scenarios, the following assumptions has been made:

- Forklifts charging operations were carried out overnight,
- Forklift charging losses were assumed to be 15% of the battery's capacity according to Rand et al. [18]
- The grid-related energy losses referred to on-site power renewable generation units (i.e., PV panels) are assumed as negligible,
- The air circulation rate for mechanical ventilation is assumed to be 0.9 Vol/h.
- Packing system requires an average electrical power of 5 kW,
- One Standard Cubic Meter (Sm^3) of natural gas used for energy, environmental and economic assessment is assumed equal to 10,69 kWh,
- In the assessment of $CO2_{eq}$ emissions, a conversion factor of 0.259 $gCO2_{eq}$/kWh for electricity and 0.233 $gCO2_{eq}$/kWh for natural gas has been adopted according to the ISPRA report [19],
- The investment cost of GW practices has been estimated using DesignBuilder database,
- In the economic assessment, the average price electricity and natural gas was assumed 0.23 €/kWh and 0.80 €/Sm3 respectively, according to ARERA [20].

In accordance with the previously highlighted framework, the base case scenario was simulated to compare results with real data collected by the company and identify any critical consumption areas. To achieve this purpose, heating-related natural gas consumption was converted to electricity in terms of kWh, so that consumption from different energy sources can be compared.

The monthly consumption breakdown by main end use type (i.e., heating, cooling, lighting, MHE, and mechanical ventilation or others) is reported in Fig. 3. Consumption results are in line with real data collected by the company, thus validating the model.

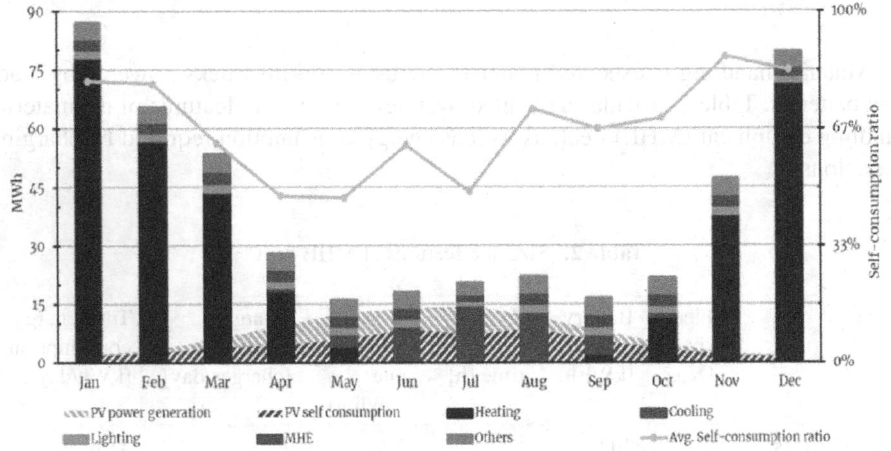

Fig. 3. Monthly electricity consumption breakdown

4.2 Scenario A

The installation of LED bulbs throughout warehouse floorspace, replacing fluorescent light bulbs (Scenario A) was first simulated. This measure can be considered a mutually beneficial solution thanks to its ease of implementation and minimal capital expenditure, as documented by GILA [21].

4.3 Scenario B

As a second step (Scenario B) the implementation of a heat pump system was simulated. This system replaced the traditional boiler system for heating generation completely electrifying the building energy load. In this context, the Coefficient of Performance (COP) for the heat pump was set equal to 2.0 for heating and 2.5 for cooling as suggested by Nyers et al. [22]. Based on yearly heating and cooling loads of the logistics facility, the heat pump system must have a power capacity of 250 kW to meet them.

4.4 Scenario C

Thermal insulation of the envelope was then implemented by setting parameters related to the building construction layers. Specifically, variations of the insulation thicknesses added to the external envelope and structural and thermal features of the openings were simulated Each intervention was estimated in compliance with the current technical standards UNI EN ISO 9946 (i.e., related to the external envelope) and UNI EN ISO 10077–1 (i.e., related to openings). Thus, transmittance values achieved are 0.22 W/m^{2}K for roofs and external walls, 0.57 W/m^{2}K for external and internal walls and 1.22 W/m^{2}K for openings.

4.5 Scenario D

Lastly, the joint adoption of both thermal insulation of the envelope insulation and heat pump system was simulated. As modifying the envelope's thermal insulation results in a significant impact on the buildings' demand profile(and thus the resulting peak load) ([23, 24]), thermal insulation was assumed to be first implemented, followed by the heat pump system's installation, so that it could be sized according to the already reduced demand profile. Accordingly, based on the new yearly heating and cooling loads of the logistics facility, the heat pump system required a power capacity of 90 kW.

5 Results and Discussion

The simulation results of Scenario A show that the logistics facility's annual electricity demand slightly decreased. Even though LED light bulbs have a notable impact on lighting consumption compared to the base case (−25%), the introduction of such GW practice has a negligible impact due to the small contribution of lighting to total energy consumption. Such measure is also extremely cost-effective, with a high Return on Investment (ROI) equal to 70%, and a quick Payback Period (PBP) of about 1 year.

Regarding Scenario B, results demonstrate that the total yearly energy demand further decreases (−53%), compared to the base case, mainly due to both energy savings in cooling (−68%) and heating (−60%). On the other hand, average monthly self-consumption ratio further increases (+5%) due to the large increase in electrical loads (i.e., electrification of heating system). Nevertheless, such a scenario is not particularly economically viable due to limited electricity savings. Although gas consumption was totally removed, electricity consumption doubled due to high building thermal loss. Looking at Scenario C, findings indicate that total yearly energy demand decreases (−34%) compared to the base case but at a lesser extent if compared to Scenario B (+41%). These results are mainly due to the inefficiency of the current HVAC system (i.e., conventional gas-fired boiler) leading to financial consequences (i.e., long payback period, limited cost savings). Finally, in Scenario D the joint implementation of thermal insulation envelope followed by heat pump system was simulated. Hence, results show a large yearly energy consumption decrease compared to the previous case (−62%) and base case (−75%) scenario. On the other hand, the average monthly self-consumption ratio was reduced compared to the base case scenario (−4%) due to the lowest value achieved in terms of annual energy consumption compared to all other scenarios investigated. To provide further insights on the impact of the investigated interventions, monthly energy consumption by energy source (electricity and natural gas), environmental KPIs – e.g., Energy Used Index (EUI), $CO2_{eq}$ emission intensity – and some relevant economic KPIs (e.g., ROI, PBP) for the examined scenario are reported in Table 3.

Table 3. Monthly energy consumption [MWh] breakdown and KPIs for each scenario.

	BASE CASE		SCENARIO A		SCENARIO B		SCENARIO C		SCENARIO D	
Months	* Net Electricity	Gas	* Net Electricity	Gas	* Net Electricity	Gas	* Net Electricity	Gas	* Net Electricity	Gas
Jan	7,1	78.3	6,6	78.7	38,3	–	7,1	46,0	17,5	–
Feb	6,0	57.0	5,6	57.5	28,6	–	6,2	35,4	13,7	–
Mar	5,7	43.7	5,4	44.1	21,9	–	6,1	28,7	11,0	–
Apr	4,8	18.6	4,6	18.9	10,6	–	5,2	13,7	6,4	–
May	5,8	4.3	5,6	4.3	5,9	–	5,3	4,0	4,5	–
Jun	8,9	0.7	8,5	0.8	5,2	–	5,9	0,9	4,1	–
Jul	13,2	0.1	12,7	0.1	4,5	–	5,2	0,1	2,7	–
Aug	13,2	0.3	12,6	0.3	6,3	–	7,6	0,3	4,8	–
Sep	8,7	2.3	8,2	2.3	6,9	–	7,3	2,3	5,7	–
Oct	6,5	11.8	6,1	12.0	10,5	–	6,8	8,4	7,2	–
Nov	7,3	38.0	6,8	38.4	22,5	–	7,7	23,7	12,4	–
Dec	6,3	71.8	5,9	72.1	34,9	–	6,3	36,3	14,5	–
Total energy consumption [MWh]	93,4	326.8	88,6	329.5	196,2	–	76,8	199,7	104,5	–
	420,2		418,1		196,2		276,5		104,5	
EUI [kWh/m^2]	120,1		119,5		56,0		79,0		29,9	
Avg. Monthly self consumption ratio	67%		65%		70%		62%		63%	
$CO2_e$ emission intensity [$kgCO2_e/m^2$]	28.5		28.3		14.3		18.8		7.6	
ROI	–		70%		6%		3%		5%	
PBP [years]	–		1.4		14		35		20	

* Net Electricity refers to the electricity required minus energy produced and self-consumed on-site from renewable energy (i.e., PV panels).

Moreover, monthly CO2eq emissions for each scenario were also computed, as reported in Fig. 4.

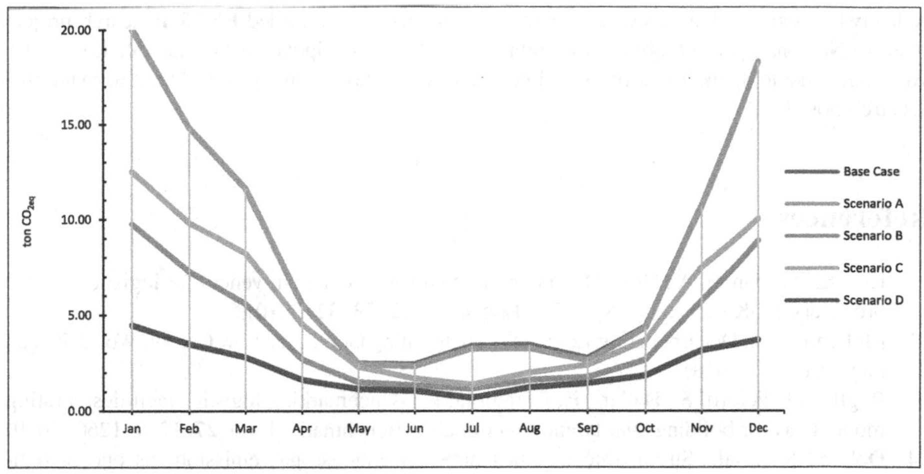

Fig. 4. Monthly CO2eq emissions for each scenario

6 Conclusions

In this study, a simulation-based approach was proposed with multiple scenarios of a real conditioned logistics facility, grounded on a conceptual framework that offers a roadmap towards sustainable warehousing. The simulation was conducted under five scenarios: existing condition, LED lighting systems implementation, heat pumps installation, thermal insulation of the envelope, and the simultaneous implementation of thermal insulation and heat pump system. Results were analyzed and discussed based on different KPIs such as monthly energy consumption by energy source, Energy Used Index (EUI), $CO2_{eq}$ emission intensity, ROI and PBP. It was demonstrated that LED light bulbs are the most cost-effective measure with a high return on investment and a short payback time in line with [10]. However, a critical correlation was noted between heat pump system's implementation and thermal insulation envelope. Even though these GW measures provide significant energy savings, they appear as not cost-effective if implemented individually. Conversely, by thermally insulating the logistics building first, heat pump size is determined based on a much lower energy demand compared to sizing on base case, avoiding potential energy inefficiencies and providing economic benefits. Although this study addresses a notable gap in the existing literature, it does come along certain limitations, since maintenance expenses of GW measures investigated were neglected. Moreover, it focused exclusively on electricity and gas consumption intentionally ignoring other sources (e.g., refrigerants, waste, water) that have typically a small impact (i.e., around 5% of the GHG emissions of ambient temperature sites in Italy can be allocated to this sources of consumption) as pointed out by Perotti & Colicchia [5]. However, this contribution can pave the way for promising developments for future investigation: (a) performing an economic analysis that considers all the benefits and costs direct and indirect of the GW measure under investigation; (b) the simulation model can be extended by considering other energy sources; (c) battery energy storage for self-consumption ratio increase can be further evaluated.

Acknowledgement. This research is part of a broader Italian funded PNRR Research project "Centro Nazionale per la Mobilità Sostenibile" (CN MOST) – Spoke 10 "Sustainable Logistics". We would like to thank all partners and companies that have actively contributed to make this research possible.

References

1. Liu, X., Mckinnon, A., Grant, D., Feng, Y.: Sources of competitiveness for logistics service providers: a UK industry perspective. Logist. Res. **2**, 23–32 (2010)
2. McKinnon, A.: Decarbonizing Logistics - Distributing Goods in a Low Carbon World. Kogan Page, London (2018)
3. Baglio, M., Perotti, S., Dallari, F., Garagiola, E.: Benchmarking logistics facilities: a rating model to assess building quality and functionality. Benchmark.: Int. J. **27**, 1239–1260 (2019)
4. Dobers, K., et al.: Sustainable logistics hubs: greenhouse gas emissions as one sustainability key performance indicator. In: Proceedings of the Transport Research Arena (TRA) Conference (2022)
5. Perotti, S., Colicchia, C.: Greening warehouses through energy efficiency and environmental impact reduction: a conceptual framework based on a systematic literature review. Int. J. Logist. Manag. (2023)
6. Bartolini, M., Bottani, E., Grosse, E.: Green warehousing: systematic literature review and bibliometric analysis. J. Clean. Prod. **226**, 242–258 (2019)
7. Freis, J., Vohlidka, P., Günthner, W.A.: Low-carbon warehousing: examining impacts of building and intra-logistics design options on energy demand and the co2 emissions of logistics centers. Sustainability **8**(5), 448 (2016)
8. Ou, Y., Wang, X., Liu, J, Warehouse multipoint temperature and humidity monitoring system design based on king view. In: 'AIP conference proceedings', vol. 1834. AIP Publishing (2017)
9. Fichtinger, J., Ries, J.M., Grosse, E.H., Baker, P.: Assessing the Environmental Impact of Integrated Inventory and Warehouse Management. Int. J. Prod. Econ. **170**, 717–729 (2015)
10. Füchtenhans, M., Glock, C.H., Grosse, E.H., Zanoni, S.: Using smart lighting systems to reduce energy costs in warehouses: a simulation study. Int. J. Log. Res. Appl. **26**(1), 77–95 (2023)
11. Maurer, W.: Light management below ground: lighting technology in tunnelling. Geomechanik und Tunnelbau **8**(4), 348–355 (2015)
12. Staffell, I., Brett, D.J.L., Brandon, N.P., Giarola, S.: A review of domestic heat pumps. Energy Environ. Sci. **5**(11), 9291 (2012)
13. Scoccia, R., Toppi, T., Aprile, M., Motta, M.: Absorption and compression heat pump systems for space heating and DHW in European buildings: energy, environmental and economic analysis. J. Build. Eng. **16**, 94–105 (2018)
14. Mishra, S.C., Usmani, D.J., Varshney, S.K.: Energy saving analysis in building walls through thermal insulation system (2012)
15. Paraschiv, S., Paraschiv, L.S., Serban, A.: Increasing the energy efficiency of a building by thermal insulation to reduce the thermal load of the micro-combined cooling, heating and power system. Energy Rep. **7**, 286–298 (2021)
16. Siems, E., Seuring, S.: Stakeholder management in sustainable supply chains: a case study of the bioenergy industry. Bus. Strateg. Environ. **30**(7), 3105–3119 (2021)
17. Cook, P., Sproul, A.: Towards low-energy retail warehouse building. Archit. Sci. Rev. **54**(3), 206–214 (2011)

18. Rand, D.A. Moseley, P.T.: Chapter 13 - energy storage with lead–acid batteries. In: Electrochemical Energy Storage for Renewable Sources and Grid Balancing, pp. 201–222. Elsevier (2015)
19. ISPRA. ISPRA Report 2022: Indicatori di efficienza e decarbonizzazione del sistema energetico nazionale e del settore elettrico (2022). www.isprambiente.gov.it. Accessed 22 Sep 2023
20. ARERA (Autorità di Regolazione per l'Energia Reti e Ambiente), prezzi finali dell'energia per i consumatori industriali – Ue a Area euro (2022). https://www.arera.it. Accessed 22 Sep 2023
21. GILA (2021). https://www.som.polimi.it/en/reducing-the-environmental-impact-of-logistics-the-gila-project/. Accessed 22 Sep 2023
22. Nyers, J.M., Nyers, Á.J.: COP of heating-cooling system with heat pump. In: EXPRES 2011 - 3rd IEEE International Symposium on Exploitation of Renewable Energy Sources, Proceedings, pp. 17–21 (2011). https://doi.org/10.1109/EXPRES.2011.5741809
23. Maltseva, I.N., Elokhov, A., Tkachuk, K., Maltceva, K.: Design without thermal bridges. In: MATEC Web of Conferences, vol. 146, p. 03002 (2018)
24. Sarevet, H., Fadejev, J., Thalfeldt, M., Kurnitski, J.: Residential buildings with heat pumps peak power reduction with high performance insulation. In: E3S Web of Conferences, vol. 172, p. 12008 (2020)

Zero-Inflated Poisson Tensor Factorization for Sparse Purchase Data in E-Commerce Markets

Keisuike Mizutani, Ayaka Ueta, Ryota Ueda, Ray Oishi, Tomofumi Hara, Yuki Hoshino, Ken Kobayashi(✉), and Kazuhide Nakata

Tokyo Institute of Technology, Tokyo, Japan
{mizutani.k.ad,ueta.a.aa,ueda.r.ae,oishi.r.ab,hara.t.ap,hoshi.y.ad, kobayashi.k.ar}@m.titech.ac.jp

Abstract. Nonnegative tensor factorization (NTF) plays a crucial role in extracting latent factors and predicting future sales from purchase data consisting of user and item attributes. However, the increase in these attributes leads to tensor data becoming sparse, causing a reduction in decomposition accuracy. For example, when there are numerous combinations of unavailable item genres and prices, the purchase history data becomes sparse and follows a distribution where all its elements are zero. To address this issue, we propose a novel NTF method assuming zero-inflated Poisson (ZIP) distribution based on Expectation-Maximization (EM) algorithm. This enables us to effectively handle sparsity in high-dimensional multiway data and identify combinations of user and item attributes that are potentially not likely to be purchased. We verified the effectiveness of the proposed approach through numerical experiments using real-world e-commerce data. The results showed our proposed ZIP model outperforms existing methods in both in-sample and out-of-sample experiments. Moreover, the proposed method qualitatively demonstrated the effectiveness of handling sparsity.

Keywords: tensor factorization · zero-inflated Poisson distribution · e-commerce

1 Introduction

E-commerce markets play a crucial role in modern online businesses, and the analysis of customer purchase data is essential for increasing sales. Such purchase data can be expressed as nonnegative multiway data, which consists of customer, product, and time period attributes, and nonnegative tensor factorization (NTF) [1] is often employed to analyze it. NTF decomposes multiway data into a product of nonnegative matrices, each corresponding to a dimension called "mode". This process not only compresses the original data but also enables us to investigate the underlying relationship among the modes or identify hidden patterns within the data. Also, as the extracted factor matrices can serve

S.-H. Sheu (Ed.): ICIEA-EU 2024, LNBIP 507, pp. 158–171, 2024.
https://doi.org/10.1007/978-3-031-58113-7_14

as latent representations in the corresponding attributes, these representations can be effectively used in various kinds of other tasks in e-commerce markets, such as recommendation and purchase prediction [2–4].

In recent years, the ease of data collection has led to an increase in the number of attributes that can serve as modes in tensor data in e-commerce [5–7]. These various attributes enable us to conduct a more detailed analysis of tensor data. However, suppose that we stratify count or gross summation data according to all available modes; the tensor data would get extremely sparse. In particular, for some combination of item and user attributes, non-existent items for example consistently exhibit a zero value (such as low-priced jewelry). This highly sparse data leads to instability and a potential decline in factorization accuracy.

To address this issue, we propose a new NTF method for high-dimensional sparse purchase data for e-commerce markets. To consider the sparsity of the original data, our method assumes that each element of the data is drawn from zero-inflated Poisson (ZIP) distributions [8]. A ZIP distribution combines a Poisson distribution which is normally used to represent count data [9] with a distribution composed entirely of zero values. This distribution has the capability of effectively expressing data containing an excess number of zero values. Abe and Yoshida [10] proposed two-dimensional nonnegative matrix factorization (NMF) assuming the zero-inflated distribution for two-dimensional count data. We extend this existing method and propose a model that allows decomposition with a nonnegative tensor with three or more dimensions as shown in Fig. 1. In addition, we design the mixture ratio of ZIP distribution along any desired modes, a departure from the existing model where it was fixed across the entire element of the matrix. This improves the accuracy of tensor factorization and enables the estimation of latent purchase probabilities at the item and user levels.

To investigate the effectiveness of our method, we conducted in-sample and purchase prediction experiments with actual purchase data from online retail stores. The results of both experiments demonstrate that the proposed method can predict the tensor data with higher accuracy than the existing methods. Furthermore, from the mixture ratio obtained by our proposed method, valuable interpretations for marketing can be facilitated such as predicting users' potential behaviors. Although this paper focuses on purchase data, our proposed approach is not limited to marketing analysis. It can also be applied to other sparse tensor data, enabling detailed analysis compared to the conventional NTF methods.

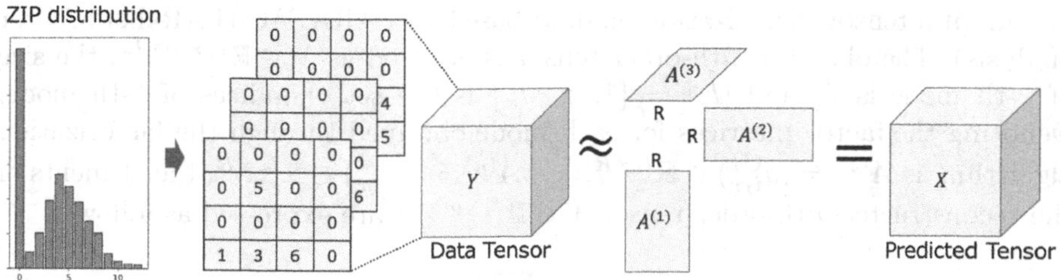

Fig. 1. Tensor factorization for sparse data

2 Related Work

2.1 Tensor Factorization

Tensor factorization is a method that decomposes high-dimensional tensor data with multiple modes into low-dimensional factors [11]. This extends matrix factorization to multiple dimensions and proves to be effective in revealing the underlying structure of the data.

Tensor factorization is widely employed in various purposes [12] such as image restoration [7,13,14], audio source separation [15], and network analysis [16]. By decomposing complex data structures into low-dimensional factors, it enables to achieve dimensionality reduction and noise removal. Consequently, this method proves to be effective in extracting data features and obtaining interpretable information. However, there are challenges associated with the high computational costs and non-uniqueness of solutions [7,17], which are critical issues to consider.

2.2 Zero-Inflated Poisson

Existing factorization algorithms, such as NMF [1], face difficulties in approximating sparse data with a high proportion of zero values. To address this issue, factorization assuming ZIP distribution and zero truncated Poisson (ZTP) distribution in data were proposed [10,18]. In factorization assuming ZTP distribution, all zero entries are ignored and its distribution is applied only on the positive counts. However, the high dimensionality and fine granularity of the data used in this study generate a significant number of zero values. Thus, we need to distinguish between true and false zero counts, making the ZIP distribution more appropriate for our model.

Abe et al. [10] proposed a new NMF model designed explicitly for zero-inflated nonnegative matrices, assuming non-zero values to follow a Tweedie distribution. In contrast, we assume that the non-zero values follow a Poisson distribution and extend the existing model to be capable of decomposing tensors with three or more dimensions. Additionally, we enable the individual specification of mixture ratios for the ZIP distribution, which were uniform across the entire matrix in the existing model.

3 Proposed Method

We adopt a tensor factorization method based on PARAFAC (PARallel FACtor analysis). The observed nth-order tensor is denoted as $\mathcal{Y} \in \mathbb{R}^{I_1 \times \cdots \times I_n}$, the size of k-th mode is I_k and $[I_k] = \{1, \ldots, I_k\}$ is the set of indices of k-th mode. Denoting the factor matrices for each mode obtained through the factorization algorithm as $\boldsymbol{A}^{(1)} = (a^{(1)}_{i_1 r}) \in \mathbb{R}^{I_1 \times R}, \ldots, \boldsymbol{A}^{(n)} = (a^{(n)}_{i_n r}) \in \mathbb{R}^{I_n \times R}$, the elements of the reconstructed nth-order tensor $\mathcal{X} \in \mathbb{R}^{I_1 \times \cdots \times I_n}$ are expressed as follows:

$$x_{i_1, \ldots, i_n} = \sum_r \prod_{l=1}^{n} a^{(l)}_{i_l r}. \tag{1}$$

Tensor factorization often employs the ALS (Alternating Least Squares) algorithm and assumes a normal distribution for the given tensor. This algorithm calculates the tensor's original form and the reconstructed tensor in such a way that the squared error between them is minimized. This is because when the squared error is minimal, the likelihood of the predictions is maximized when assuming a normal distribution. However, we have assumed a ZIP distribution for our data, so the likelihood function to be maximized is different from them. This section describes an updated tensor factorization formula based on the proposed method for the ZIP distributions.

The purchase prediction framework is illustrated in Fig. 2 (b). We adopt a method utilizing the time series factor of decomposed tensor proposed in existing research [19]. 'Week' data were used as the time series factor. The experimental results of this framework are summarised in Sect. 4.3.

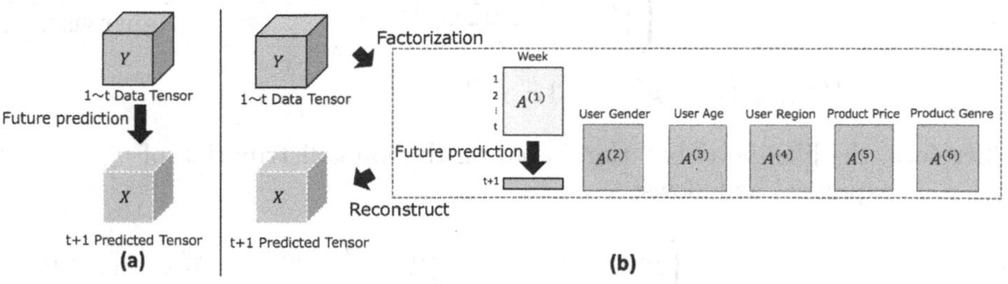

Fig. 2. Purchase prediction frameworks; (a) Element-wise purchase prediction model. (b)Tensor factorization model assuming each distribution. Subsequently, it predicts the Week factor using a prediction model, then reconstructs the predicted tensor.

3.1 ZIP Distribution Model

A probability variable $y \sim \text{ZIP}(\lambda, p)$ is characterized by a mixture distribution consisting of only zero counts and a Poisson distribution. Here, p is the mixture ratio of ZIP distribution and λ is the expected value of the Poisson distribution [8]. The probability mass function of ZIP distribution is defined as follows:

$$f(y; \lambda, p) = \begin{cases} p + (1-p)\exp(-\lambda) & \text{if } y = 0, \\ (1-p)\frac{\lambda^y \exp(-\lambda)}{y!} & \text{otherwise.} \end{cases} \tag{2}$$

Now, let us consider an nth-order tensor $\mathcal{Y} \in \mathbb{R}^{I_1 \times \cdots \times I_n}$. An tensor element located at position $(i_1, \ldots, i_n)$ in the tensor is denoted as $y_{i_1 \ldots i_n}$, its predicted value is denotd as $x_{i_1 \ldots i_n}$, and $y_{i_1 \ldots i_n}$ follows a ZIP distribution. The likelihood function for an element $y_{i_1 \ldots i_n}$ with respect to the parameters p and $x_{i_1 \ldots i_n}$ (the predicted value) is

$$L(y_{i_1 \ldots i_n}; x_{i_1 \ldots i_n}, p) = \begin{cases} p + (1-p)\exp(-x_{i_1 \ldots i_n}) & \text{if } y_{i_1 \ldots i_n} = 0, \\ (1-p)\frac{\exp(-x_{i_1 \ldots i_n}) x_{i_1 \ldots i_n}^{y_{i_1 \ldots i_n}}}{y_{i_1 \ldots i_n}!} & \text{otherwise.} \end{cases}$$

We use the Expectation-Maximization (EM) algorithm [20] to estimate the parameters of this mixture distribution for maximizing likelihood. In order to perform estimation using the EM algorithm, the following latent variables are introduced:

$$z_{i_1\dots i_n} = \begin{cases} 1 & \text{if } y_{i_1\dots i_n} \sim 0, \\ 0 & \text{if } y_{i_1\dots i_n} \sim \text{Poisson}\left(x_{i_1\dots i_n}\right). \end{cases}$$

Based on the definition of latent variables, the distribution of $y_{i_1\dots i_n}$ when $z_{i_1\dots i_n}$ is obtained, and the distribution of $z_{i_1\dots i_n}$ which is assumed to follow a Bernoulli distribution, are given as follows:

$$f\left(y_{i_1\dots i_n}; z_{i_1\dots i_n}, x_{i_1\dots i_n}\right) = \begin{cases} 1^{z_{i_1\dots i_n}} \left[\dfrac{e^{-x_{i_1\dots i_n}} x_{i_1\dots i_n}^{y_{i_1\dots i_n}}}{y_{i_1\dots i_n}!}\right]^{1-z_{i_1\dots i_n}} & \text{if } y_{i_1\dots i_n} = 0, \\ \left[\dfrac{e^{-x_{i_1\dots i_n}} x_{i_1\dots i_n}^{y_{i_1\dots i_n}}}{y_{i_1\dots i_n}!}\right]^{1-z_{i_1\dots i_n}} & \text{otherwise}, \end{cases}$$

$$f(z_{i_1\dots i_n}; p) = p^{z_{i_1\dots i_n}} (1-p)^{(1-z_{i_1\dots i_n})}.$$

Therefore, the likelihood function based on the joint distribution of $y_{i_1\dots i_n}$ and $z_{i_1\dots i_n}$ is expressed as follows:

$$L\left(y_{i_1\dots i_n}; x_{i_1\dots i_n}, p\right) = \begin{cases} p^{\hat{z}_{i_1\dots i_n}} \left[(1-p)e^{-x_{i_1\dots i_n}}\right]^{1-\hat{z}_{i_1\dots i_n}} & \text{if } y_{i_1\dots i_n} = 0, \\ \left[(1-p)\dfrac{e^{-x_{i_1\dots i_n}} x_{i_1\dots i_n}^{y_{i_1\dots i_n}}}{y_{i_1\dots i_n}!}\right]^{1-\hat{z}_{i_1\dots i_n}} & \text{otherwise}. \end{cases}$$

Here, $\hat{z}_{i_1\dots i_n} \in [0,1]$ represents the estimated values of the latent variables. Then, the sum of the log-likelihoods (LL) of all elements of the tensor is given as follows:

$$\begin{aligned} &\sum_{i_1\dots i_n} LL\left(y_{i_1\dots i_n}; x_{i_1\dots i_n}, p\right) \\ &= \sum_{i_1\dots i_n} \left(\hat{z}_{i_1\dots i_n} \log p + (1-\hat{z}_{i_1\dots i_n}) \log\left((1-p)\frac{\exp(-x_{i_1\dots i_n}) x_{i_1\dots i_n}^{y_{i_1\dots i_n}}}{y_{i_1\dots i_n}!}\right)\right). \end{aligned} \tag{3}$$

Although the formulation is presented for an arbitrary nth-order tensor, in the numerical experiments of Sect. 4, we used a sixth-order tensor. In this case, the algorithm for tensor factorization is designed to solve the following minimization problem:

$$\begin{aligned} \min_{\boldsymbol{A}^{(1)},\dots,\boldsymbol{A}^{(n)}} \quad & -\sum_{i_1\dots i_n} LL\left(y_{i_1\dots i_n}; x_{i_1\dots i_n}, p\right) \\ \text{s.t.} \quad & x_{i_1\dots i_n} = \sum_r \prod_{l=1}^{n} a_{i_l r}^{(l)} \quad (\forall (i_1, \dots, i_n) \in [I_1] \times \dots \times [I_n]), \\ & \boldsymbol{A}^{(1)}, \dots, \boldsymbol{A}^{(n)} \geq 0. \end{aligned}$$

$\boldsymbol{A}^{(1)}, \ldots, \boldsymbol{A}^{(n)}$ represent the factor matrices after factorization for each mode, and R is the rank. Since the data is assumed to follow a ZIP distribution, nonnegativity constraints are imposed on each factor matrix. In the E-step, the latent variables $z_{i_1 \ldots i_n}$ is updated using the conditional expected value given $y_{i_1 \ldots i_n}$:

$$\hat{z}_{i_1 \ldots i_n} = \begin{cases} \frac{p}{p+(1-p)\exp(-x_{i_1 \ldots i_n})} & \text{if } y_{i_1 \ldots i_n} = 0, \\ 0 & \text{otherwise.} \end{cases} \tag{4}$$

In the M-step, we first update p using the latent probability $z_{i_1 \ldots i_n}$. The update equation for the mixture ratio that maximizes LL is

$$p = \frac{1}{I_1 \cdots I_n} \sum_{i_1 \ldots i_n} z_{i_1 \ldots i_n}. \tag{5}$$

The updates for $\boldsymbol{A}^{(1)}, \ldots, \boldsymbol{A}^{(n)}$ are obtained by solving the above minimization problem. From Jensen's inequality and auxiliary variable [1], the following equation holds using only the terms that include elements of factor matrices from Eq. (3):

$$\sum_{i_1 \ldots i_n} \left((1 - \hat{z}_{i_1 \ldots i_n}) \sum_r \phi_{i_1 \ldots i_n r} + (1 - \hat{z}_{i_1 \ldots i_n}) y_{i_1 \ldots i_n} \log \sum_r \left(\theta_{i_1 \ldots i_n r} \frac{\phi_{i_1 \ldots i_n r}}{\theta_{i_1 \ldots i_n r}} \right) \right)$$

$$\geq \sum_{i_1 \ldots i_n} \left((1 - \hat{z}_{i_1 \ldots i_n}) \sum_r \phi_{i_1 \ldots i_n r} + (1 - \hat{z}_{i_1 \ldots i_n}) y_{i_1 \ldots i_n} \sum_r \theta_{i_1 \ldots i_n r} \log \frac{\phi_{i_1 \ldots i_n r}}{\theta_{i_1 \ldots i_n r}} \right)$$

$$\text{where } \phi_{i_1 \ldots i_n r} = \prod_{l=1}^{n} a_{i_l r}^{(l)}.$$

Here, $\theta_{i_1 \ldots i_n r}$ is an auxiliary variable. Estimating the auxiliary variable from the equality condition, and differentiating the element $a_{i_k r}^{(k)}$ at position (i_k, r) of the factor matrix for mode k, the following update equation is given as follows:

$$a_{i_k r}^{(k)} = \frac{\sum\limits_{i_1, \ldots, i_{k-1}, i_{k+1}, \ldots, i_n} (1 - z_{i_1 \ldots i_n}) y_{i_1 \ldots i_n} \theta_{i_1 \ldots i_n r}}{\sum\limits_{i_1, \ldots, i_{k-1}, i_{k+1}, \ldots, i_n} (1 - z_{i_1 \ldots i_n}) \prod\limits_{\substack{l=1 \\ l \neq k}}^{n} a_{i_l r}^{(l)}}, \tag{6}$$

$$\text{where } \theta_{i_1 \ldots i_n r} = \left(\prod_{l=1}^{n} a_{i_l r}^{(l)} \right) \Big/ \left(\sum_r \prod_{l=1}^{n} a_{i_l r}^{(l)} \right). \tag{7}$$

3.2 Mixture Ratio Individualization

In the existing method [10], the mixture ratio is uniformly maintained for elements across the entire matrix. In contrast, we propose the mixture ratios be uniform across arbitrary modes. Let $\mathcal{P} = (p_{i_1 \ldots i_n}) \in \mathbb{R}^{I_1 \times \cdots \times I_n}$ represent the mixture ratios defined for each element, and let the set of mode numbers for which the mixture ratio is unified be denoted as $Q = \{q_1, \dots, q_m\}$ $(1 \leq q_1 < \dots < q_m \leq n)$. Then, the Eq. (5) can be expressed as:

$$p_{i_1,\dots,i_n} = \frac{1}{I_{q_1} \cdots I_{q_m}} \sum_{i_{q_1} \dots i_{q_m}} z_{i_1 \dots i_n}. \tag{8}$$

When $Q = \{1, ..., n\}$, the proposed model aligns with the existing method. Based on the update formula and assuming a ZIP distribution, the proposed factorization algorithm is shown in Algorithm 1.

Algorithm 1. Tensor factorization algorithm with ZIP distribution

```
Input: Observed tensor 𝒴, number of ranks R
Initialize: A^(1), ..., A^(n), 𝒫
while Convergence conditions not satisfied. do
    The EM algorithm is repeated below.
    Update the latent variables ẑ using Equation (4)
    Update the mixture ratios p using Equation (8)
    Update each factor matrix A^(1), ..., A^(n) using Equation (6)
end while
Output: A^(1), ..., A^(n), 𝒫
```

4 Experiments and Discussion

We conducted numerical experiments to verify our proposed tensor factorization algorithm for both in-sample and out-sample data. To ensure the high-speed performance of our proposed tensor factorization algorithm, we provided a practical implementation using JAX, a specialized library for automating differentiation, and the Just In Time (JIT) compiler. In this implementation, we are able to perform factorization within the computational time constraints, without relying on an increase in the number of factors, in contrast to the existing factorization library, Tensorly [21] (Appendix A). Our implemented algorithms for multi-mode tensors can also be applied more generally beyond the scope of this research. They have been open-sourced and are available on GitHub[1].

4.1 Data Sets and Evaluation Indicators

The data set was provided by[2]Rakuten Ichiba to illustrate the applicability of our proposed algorithm to actual purchase history data. The data set consists of 12,888,912 purchase records of 100,000 randomly selected individuals for each

[1] https://github.com/tokyotech-nakatalab/tensor-decomposition.git.
[2] https://www.rakuten.com/.

age group from 2019 to 2020. An overview of each column is presented in Table 1. In the following experiments, we constructed a sixth-order tensor based on these columns. This tensor was highly sparse with over 96% of its elements being zero, making our proposed algorithm more suitable.

Table 1. Data set overview

Column	Num Unique	Description	Examples
week	104	week ID	1–104
user_gender	2	User Gender	m = Male, f = Female
user_age	4	User Age	(20, 35], ..., (65, 80]
region	8	User Region	Hokkaido, ..., Kyushu
price	31	Product Price	(0, 1k], ..., (3 m, 4 m]
genre_name	531	Product Genre	Shoes > ··· > Sneakers

Regarding the necessity of considering differences in scale, we used different evaluation metrics for zero and non-zero values of the observed data: Route Mean Squared Percentage Error (RMSPE) for non-zero values and RMSE, ROC-AUC, and PR-AUC for zero values. For RMSPE and RMSE, a smaller value indicates better accuracy while a larger value indicates better performance for the others.

4.2 Comparative Analysis of Factorization Algorithms

We conducted preliminary experiments to determine the dependency between the predicted values and the initial decomposed values. The results showed that factorization assuming a normal distribution and ZIP distribution has no dependency on initial values, allowing relatively stable factorization and reconstruction. However, the factorization algorithm assuming a Poisson distribution relies heavily on initial values. Thus, we designed its algorithm to use the decomposed results of a normal distribution as the initial values. The tensor factorization models employed in all cases incorporate a non-negative constraint.

In this experiment, we calculated evaluation metrics for the original tensor and its decomposed and reconstructed tensor using in-sample data. We generated a predicted average tensor where each element represents the average values from 10 trials of factorization and reconstruction process with $R = 5$. Table 2 shows the comparison of three factorization algorithms, which each represent a normal distribution (Normal), Poisson distribution (Poisson), and ZIP distribution (ZIP). For the ZIP distribution, we implemented two algorithms that define the mixture ratio p common across all elements ($Q = \{1, \dots, 6\}$) used in the existing method, and that can vary per element. We diversified mixture ratios for genre (mode 5) and price (mode 6) combinations on a per-element basis ($Q = \{1, \dots, 4\}$) . The results show that our proposed algorithm outperformed existing algorithms in terms of all evaluation metrics.

Table 2. Evaluation metrics for each method

Distribution	Non-Zeros	Zeros		
	RMSPE	RMSE	ROC-AUC	PR-AUC
Normal	0.9313	0.1836	0.9662	0.6295
Poisson	0.9684	0.1877	0.9731	0.6656
ZIP($Q = \{1, \ldots, 6\}$)	0.9106	0.1820	0.9748	0.6705
ZIP($Q = \{1, \ldots, 4\}$)	**0.8869**	**0.1691**	**0.9782**	**0.6990**

4.3 Comparative Analysis of Prediction Models

Next, we conducted experiments using out-of-sample data to verify the accuracy of our proposed algorithm within the framework of purchase prediction shown in Fig. 2 (b). To conduct a comparative analysis of factorization methods, we employed a basic linear regression (LR) as the purchase prediction model.

We compared the following six models in evaluating the prediction accuracy of our proposed algorithm within this framework. We set two baseline models, one that does not employ factorization or prediction (Baseline) and one that employs only prediction (LR). The former model simply uses the data of the previous week as the predicted values and the latter performs prediction per element of a tensor as shown in Fig. 2 (a). In models using tensor factorization, we factorized the data tensor into each factor as shown in Fig. 2 (b). Then, we reconstructed the tensor using the predicted time series factor and the other inherited factors to predict the future. Four tensor factorization models were compared which assume a normal distribution (LR+Normal), Poisson distribution (LR+Poisson), and ZIP distribution (LR+ZIP). For the model assuming a ZIP distribution, we compared two variations: one that defines the mixture ratio p as a scalar that is common among all elements ($Q = \{1, \ldots, 6\}$) and one defining element-wise mixture ratio ($Q = \{1, \ldots, 4\}$).

The training data consisted of weekly purchase counts from January 2019 to the week prior to the target week, and prediction data were the weekly counts for the last two months of 2020. To determine the optimal number of factors for factorization, we conducted a preliminary experiment. Here, we used the data from the final week of train data and verified the accuracy of purchase prediction using the proposed algorithm based on RMSE and RMSPE. This plot is shown in Fig. 3 for each rank R. Regarding the decrease in interpretability of each factor for a large rank, we selected $R = 31$ for the subsequent experiments, which showed relatively small values for both metrics.

Table 3 presents the average values of the evaluation metrics for the 8 weeks of test data. From this result, it is evident that our proposed model had the best accuracy across all metrics. Even examining each target week individually revealed that our proposed model consistently exhibited the best performance.

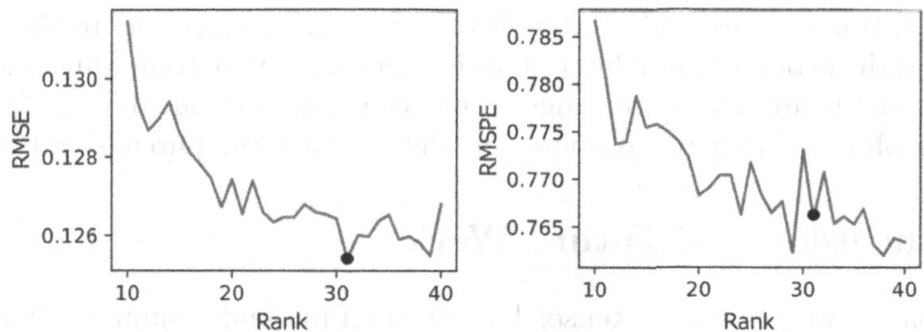

Fig. 3. Preliminary experiment for determining rank R

Table 3. Average values of evaluation metrics for each model

Prediction	Non-Zeros	Zeros		
Model	RMSPE	RMSE	ROC-AUC	PR-AUC
Baseline	1.1487	0.2632	0.7731	0.6027
LR	0.7816	0.1371	0.9432	0.7127
LR+Normal	0.7656	0.1357	0.9711	0.7122
LR+Poisson	0.7653	0.1265	0.9748	0.7286
LR+ZIP($Q = \{1, \ldots, 6\}$)	0.7495	0.1242	0.9749	0.7282
LR+ZIP($Q = \{1, \ldots, 4\}$)	**0.7480**	**0.1219**	**0.9760**	**0.7349**

4.4 Qualitative Evaluation of the Mixture Ratio

Figure 4 shows the heatmap of mixture ratios for each combination of genre and price estimated in the experiment conducted in the previous section. Areas in blue represent a higher probability of generating zero counts. From the overall result, the products in a high price range have a higher probability of resulting in zero value, making transactions less likely to occur. In typical e-commerce

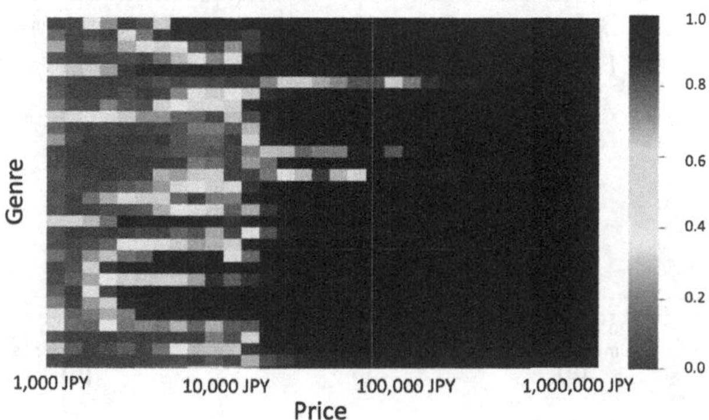

Fig. 4. Mixture ratio of ZIP distribution based on price and genre combination

platforms, transactions involving high-priced items are rare due to the absence of such products or a low number of purchasing users. Also, comparing results per each product genre, the price range of probable transactions differs. Therefore, it can be observed that our proposed model captures the purchasing behavior.

5 Conclusion and Future Work

In this paper, we proposed a tensor factorization method assuming a ZIP distribution to address the sparsity of purchase data. Unlike conventional approaches, our study extended from matrices to tensors and introduced variations in the mixture ratio of each element. The proposed method was theoretically examined and empirically validated using Rakuten's data, demonstrating improved accuracy in predicting product purchases and enabling interpretation.

Going forward, it is expected that assessing the resilience and scalability of the introduced approach will contribute to reinforcing its overall robustness. This step will involve thorough testing across diverse scenarios to ensure its adaptability to various contexts.

A Computational Time

We compare the seconds per step of the above experiments to evaluate computational time. Figure 5 (a) shows the average computational time of 10 trial runs for the four factorization models. The initial values are uniformly sampled across all models for each trial. Due to the complexity of parameters and update algorithms, our proposed method resulted in the slowest computational time.

Additionally, we compared the computational time of implementations using our JAX + JIT compiler method with Tensorly in Fig. 5 (b). Both of these methods assume a normal distribution on the tensor and use the same algorithm. Our proposed algorithm significantly mitigates the increase in computational time with the increase in rank compared to the alternative.

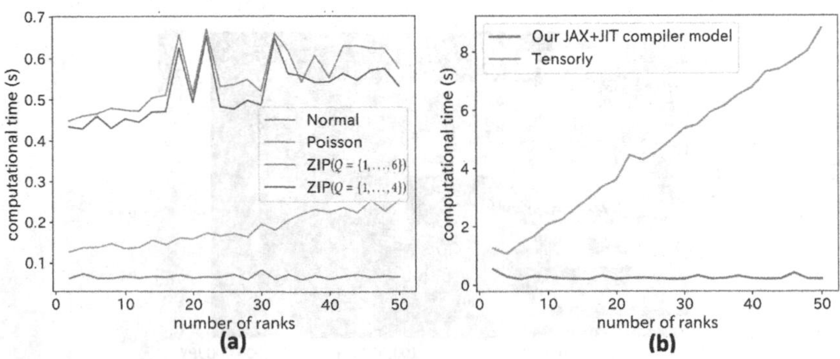

Fig. 5. Computational time per step; (a) Each tensor factorization model. (b) Backends of implementation.

B ZICMP Distribution Model

We demonstrate the Zero-Inflated Conway-Maxwell-Poisson (ZICMP) distribution tensor factorization model, which is a modified version of the Conway-Maxwell-Poisson (CMP) distribution in place of the Poisson distribution [22–24]. This enables to represent over and under dispersion which was not able by the ZIP distribution. When random variable y follows this distribution, the probability mass function is defined as follows:

$$f(y;\lambda,\nu,p) = \begin{cases} p + (1-p)\frac{1}{Z(\lambda,\nu)} & \text{if } y = 0, \\ (1-p)\frac{\lambda^y}{(y!)^\nu}\frac{1}{Z(\lambda,\nu)} & \text{otherwise,} \end{cases} \quad \text{where } Z(\lambda,\nu) = \sum_{h=0}^{\infty}\frac{\lambda^h}{(h!)^\nu}.$$

p is the mixture ratio and λ is the expected value of the ZICMP distribution. ν is a nonnegative real number and controls the variance. For $\nu = 1$, its distribution aligns with the Poisson distribution.

In contrast to the ZIP model, here we update $\boldsymbol{A}^{(1)},\ldots,\boldsymbol{A}^{(n)}$ and ν using Newton's method at each step of factor matrices updating phase in Algorithm 1. A magnitude of Newton step can be derived by taking the first and second derivatives of the log-likelihood function with respect to $a_{i_k r}^{(k)}$ and ν.

$$\Delta a_{i_k r}^{(k)} = \frac{-\sum\limits_{i_1,\ldots,i_{k-1},i_{k+1},\ldots,i_n}(1-z_{i_1\ldots i_n})\left(-\frac{Z'}{Z}+\frac{y_{i_1\ldots i_n}}{\lambda}\right)\prod\limits_{\substack{l=1\\l\neq k}}^{n}a_{i_l r}^{(l)}}{\sum\limits_{i_1,\ldots,i_{k-1},i_{k+1},\ldots,i_n}(1-z_{i_1\ldots i_n})\left(-\frac{Z''}{Z}+\left(\frac{Z'}{Z}\right)^2-\frac{y_{i_1\ldots i_n}}{\lambda^2}\right)\prod\limits_{\substack{l=1\\l\neq k}}^{n}(a_{i_l r}^{(l)})^2},$$

$$\Delta\nu = -\frac{\sum\limits_{i_1\ldots i_n}(1-z_{i_1\ldots i_n})\left(-\frac{\frac{\partial Z}{\partial\nu}}{Z}-\log(y_{i_1\ldots i_n}!)\right)}{\sum\limits_{i_1\ldots i_n}(1-z_{i_1\ldots i_n})\left(-\frac{\frac{\partial^2 Z}{\partial\nu^2}}{Z}+\left(\frac{\frac{\partial Z}{\partial\nu}}{Z}\right)^2\right)},$$

where $Z' = \sum_{h=0}^{\infty}\frac{h\lambda^{h-1}}{(h!)^\nu}$, $Z'' = \sum_{h=0}^{\infty}\frac{h(h-1)\lambda^{h-2}}{(h!)^\nu}$, $\lambda = \sum_r\prod_{l=1}^{n}a_{i_l r}^{(l)}$.

Table 4. Evaluation metrics for each algorithm

Factorization	Non-Zeros	Zeros		
Algorithm	RMSPE	RMSE	ROC-AUC	PR-AUC
Poisson	0.8186	**0.1001**	0.8484	0.6153
ZIP	0.6894	0.2101	**0.9666**	**0.7882**
ZICMP	**0.6536**	0.2836	0.9294	0.6718

We present the results of the numerical experiments for tensor factorization assuming the ZICMP distribution in Table 4. Our algorithm encountered overflow with large tensor sizes or element values. To address this, we reduced the tensor size to a 5th-mode tensor by eliminating the region mode.

As a result, RMSPE outperformed the other algorithms, but not in other evaluation metrics. While it can be assumed that the ZICMP distribution improved the expressiveness of the distribution, the fitting to sparsity did not work.

References

1. Lee, D., Seung, H.S.: Algorithms for non-negative matrix factorization. In: Advances in Neural Information Processing Systems, vol. 13. MIT Press (2000)
2. Rafailidis, D., Daras, P.: The TFC model: tensor factorization and tag clustering for item recommendation in social tagging systems. IEEE Trans. Syst., Man Cybern.: Syst. **43**(3), 673–688 (2012)
3. Taneja, A., Arora, A.: Cross domain recommendation using multidimensional tensor factorization. Expert Syst. Appl. **92**, 304–316 (2018)
4. Yin, H., et al.: SPTF: a scalable probabilistic tensor factorization model for semantic-aware behavior prediction. In: 2017 IEEE International Conference on Data Mining (ICDM), pp. 585–594. IEEE (2017)
5. Chou, S.-Y., Jang, J.-S.R., Yang, Y.-H.: Fast tensor factorization for large-scale context-aware recommendation from implicit feedback. IEEE Trans. Big Data **6**(1), 201–208 (2018)
6. Narita, A., Hayashi, K., Tomioka, R., Kashima, H.: Tensor factorization using auxiliary information. Data Min. Knowl. Disc. **25**, 298–324 (2012)
7. Zhou, P., Lu, C., Lin, Z., Zhang, C.: Tensor factorization for low-rank tensor completion. IEEE Trans. Image Process. **27**(3), 1152–1163 (2017)
8. Lambert, D.: Zero-inflated Poisson regression, with an application to defects in manufacturing. Technometrics **34**(1), 1–14 (1992)
9. Coxe, S., West, S.G., Aiken, L.S.: The analysis of count data: a gentle introduction to Poisson regression and its alternatives. J. Pers. Assess. **91**(2), 121–136 (2009)
10. Hiroyasu, A., Hiroshi, Y.: A non-negative matrix factorization model based on the zero-inflated Tweedie distribution. Comput. Statistics **32**(2), 475–499 (2017)
11. Cichocki, A., Rafal, Z., Amari, S.: Nonnegative matrix and tensor factorization [lecture notes]. IEEE Signal Process. Mag. **25**(1), 142–145 (2008)
12. Mørup, M.: Applications of tensor (multiway array) factorizations and decompositions in data mining. Wiley Interdisc. Rev.: Data Min. Knowl. Discovery **1**(1), 24–40 (2011)
13. Chen, Y., He, W., Yokoya, N., Huang, T.-Z.: Hyperspectral image restoration using weighted group sparsity-regularized low-rank tensor decomposition. IEEE Trans. Cybern. **50**(8), 3556–3570 (2020)
14. Li, S., Dian, R., Fang, L., Bioucas-Dias, J.M.: Fusing hyperspectral and multispectral images via coupled sparse tensor factorization. IEEE Trans. Image Process. **27**(8), 4118–4130 (2018)
15. Mitsufuji, Y., Roebel, A.: Sound source separation based on non-negative tensor factorization incorporating spatial cue as prior knowledge. In: 2013 IEEE International Conference on Acoustics, Speech and Signal Processing, pp. 71–75 (2013)

16. Gligorijević, V., Panagakis, Y., Zafeiriou, S.: Non-negative matrix factorizations for multiplex network analysis. IEEE Trans. Pattern Anal. Mach. Intell. **41**(4), 928–940 (2018)
17. Jain, P., Oh, S.: Provable tensor factorization with missing data. In: Advances in Neural Information Processing Systems, vol. 27. Curran Associates, Inc., (2014)
18. Lopez, O., Lehoucq, R., Dunlavy, D.: Zero-truncated Poisson tensor decomposition for sparse count data. Sandia National Lab.(SNL-NM), Albuquerque, NM (United States), Technical Report (2022)
19. Yu, H.F., Rao, N., Dhillon, I.S.: Temporal regularized matrix factorization for high-dimensional time series prediction. In: Advances in Neural Information Processing Systems, vol. 29. Curran Associates, Inc., (2016)
20. Dempster, A.P., Laird, N.M., Rubin, D.B.: Maximum likelihood from incomplete data via the EM algorithm. J. Roy. Stat. Soc.: Ser. B (Methodological) **39**(10), 1–22 (1977)
21. Kossaifi, J., Panagakis, Y., Anandkumar, A., Pantic, M.: TensorLy: tensor learning in python. J. Mach. Learn. Res. **20**, 1–6 (2016)
22. Daly, F., Gaunt, R.E.: The Conway-Maxwell-Poisson distribution: distributional theory and approximation. ALEA **13**, 635–658 (2016)
23. Gladys, F.L., Barriga, D.C.: The zero-inflated Conway-Maxwell-Poisson distribution: Bayesian inference, regression modeling and influence diagnostic. Stat. Methodol. **21**, 23–34 (2014)
24. Shmueli, G., Minka, T.P., Kadane, J.B., Borle, S., Boatwright, P.: A useful distribution for fitting discrete data: revival of the Conway-Maxwell-Poisson distribution. J. Roy. Stat. Soc.: Ser. C: Appl. Stat. **54**(1), 127–142 (2005)

Hybrid Predictive Modeling for Automotive After-Sales Pricing: Integrating BiLSTM-Attention and Fuzzy Logic

Asmae Amellal[1(✉)], Issam Amellal[2], and Mohammed Rida Ech-charrat[1]

[1] ENSA Tétouan, Abdelmalek Essaadi University, Tétouan, Morocco
asmae.amellal57@gmail.com

[2] ENSA Berrechid, Hassan 1st Univesity, Settat, Morocco

Abstract. In the automotive service and spare parts distribution sector, effective supply chain management is paramount as it directly impacts customer satisfaction, profits, and overall competitiveness. To optimize these crucial aspects, a robust fuzzy logic framework has been developed, taking into account the intricate connections between customer demand, customer sentiment, product availability, and pricing strategy. This study focuses on after-sales services provided by Moroccan automobile companies, utilizing workshop entry data from Enterprise Resource Planning (ERP) systems across multiple cities in the country.

To accurately capture customer sentiment, a Bidirectional Long Short-Term Memory (BiLSTM) with attention model, a deep learning strategy, is employed to complete any missing customer sentiment data. Leveraging the power of neural networks, this approach effectively analyzes and predicts sentiment patterns. Once the sentiment data is completed, a Fuzzy Logic model is utilized to determine the most optimal pricing strategy. By considering various factors through fuzzy sets and rules, this model allows informed decision-making to strike the right balance between profitability and customer satisfaction.

The employed models demonstrated strong performance, as evidenced by metrics such as MSE and R2. Specifically, the BiLSTM-attention model achieved an R2 of 0.87, while the fuzzy logic model registered an impressive R2 of 0.91. Furthermore, when juxtaposed with alternative models, our framework's superior efficacy became evident.

Keywords: Supply chain management · price prediction · Deep learning · BiLSTM-attention · Fuzzy method

1 Introduction

In today's increasingly competitive automotive service and spare parts distribution sector, the efficiency and effectiveness of supply chain management have become indispensable. The industry, characterized by its fast-paced nature and the ever-evolving demands of consumers, necessitates a keen understanding of various influencing factors to remain profitable and customer-centric [1]. Central to this challenge is the determination of optimal pricing strategies, which, if misjudged, can significantly affect both profitability and customer satisfaction.

S.-H. Sheu (Ed.): ICIEA-EU 2024, LNBIP 507, pp. 172–188, 2024.
https://doi.org/10.1007/978-3-031-58113-7_15

This need for accurate pricing strategies aligns well with the evolution of predictive modeling, a field that has achieved remarkable breakthroughs across a multitude of industries, opening doors to previously unattainable insights into future trends and behaviors. In the arena of sports analytics, for example, the deployment of Bayesian Networks has revolutionized the predictive accuracy, outshining traditional models in forecasting football match results with remarkable precision [2]. This not only demonstrates the robustness of these networks in handling complex datasets but also their adaptability to dynamic and unpredictable environments. Likewise, the field of health informatics has been transformed by the integration of Least Squares Support Vector Machines (LS-SVM). This advanced modeling technique has significantly outperformed conventional Neural Networks in predicting the incidence of dengue outbreaks, delivering predictions with higher accuracy and improved computational efficiency [3]. These groundbreaking applications in disparate fields underscore the substantial influence and versatility of predictive modeling, affirming its capacity to effectively navigate and elucidate the intricacies of dynamic and complex systems.

In the automotive sector, particularly in the realm of spare parts distribution, pricing strategies entail unique challenges distinct from those in manufacturing. While the pricing of manufactured automotive products is directly influenced by manufacturing costs, encompassing both fixed and variable expenses, the distribution sector operates on a different dynamic. Distributors must consider the acquisition cost, which is inherently tied to the manufacturing cost, but they also face additional layers of complexity introduced by factors such as stock conditions, distribution logistics, and customer demand. These aspects necessitate a strategic balance in pricing, ensuring that the selling price not only covers the acquisition and operational costs but also aligns with market competitiveness and customer expectations.

This complexity is further amplified by the occasional absence or inconsistency of customer sentiment data, a crucial factor in understanding market demand and shaping pricing strategies. Such data gaps or inconsistencies can significantly hinder the ability of businesses in the automotive distribution sector to fully utilize their analytical models for effective pricing [4]. Accurate interpretation and integration of customer sentiment data are vital for forecasting demand and adjusting pricing strategies accordingly.

To adeptly navigate these multifaceted challenges, our study proposes the integration of advanced deep learning strategies, specifically Bidirectional Long Short-Term Memory (BiLSTM) with attention models. BiLSTM is adept at identifying complex patterns in extensive datasets, including those scenarios where customer sentiment data is sparse or inconsistent, making it a powerful tool for analyzing market trends and demand fluctuations [5]. However, understanding customer sentiments and market trends is just one facet of the pricing strategy. The application of Fuzzy Logic models complements this by enabling nuanced decision-making that considers multiple variables simultaneously. In the context of automotive spare parts distribution, Fuzzy Logic models are invaluable in synthesizing factors like customer demand, sentiment, product availability, and acquisition costs to develop pricing strategies that are both profitable for businesses and acceptable to customers ([6, 7]).

In summary, this study aims to bridge the gap between advanced data analytics and strategic decision-making in the automotive distribution sector. By leveraging the strengths of both BiLSTM and Fuzzy Logic, it addresses the unique challenge of devising optimal pricing strategies that consider both the cost-driven constraints of distributed automotive products and the dynamic market demands.

2 Related Work

In this section, we delve into the literature surrounding the BiLSTM-Attention model for missing value predictions and the use of fuzzy logic in pricing strategies. We highlight their advantages over traditional approaches, drawing insights from recent studies in their respective fields.

2.1 BiLSTM-Attention Model

The advancement of machine learning in the field of forecasting has undeniably opened up new horizons. Over time, foundational algorithms, such as random forests and lasso, have paved the way for more sophisticated models like GRU, LSTM and BiLSTM [8]; ([9, 10]). This evolution is further enriched by hybrid models like CNN-LSTM, which [11] and others have explored.

Another game-changer in this landscape is the attention mechanism. This innovative tool acts akin to a spotlight, highlighting essential data segments, allowing models to derive more profound insights. Such a mechanism has set unprecedented benchmarks in areas like text analysis and image processing [12].

However, despite these advancements, the literature often falls short in addressing the practical applicability and limitations of these models. For instance, while complex models are celebrated for their accuracy, they sometimes come with increased computational costs and challenges in interpretability.

The BiLSTM-Attention model emerges as a balanced solution amidst this backdrop. By combining the bidirectional capabilities of BiLSTM with the pinpoint focus of attention mechanisms, it offers a pragmatic approach to various challenges. [13] and [14] have aptly demonstrated its versatility in areas ranging from health to energy forecasting.

In our endeavor, we harness the BiLSTM-Attention model specifically to address missing values in datasets. While it retains the sophistication of advanced models, it does not bear the same complexity or computational heft as models like CNN-LSTM. This ensures an efficient, yet accurate approach. However, it is worth noting that while the attention mechanism enhances accuracy, it requires a meticulous calibration to avoid over-reliance on certain data points. Our study aims not just to leverage this model but also to contribute a nuanced understanding of its applicability and potential pitfalls, filling a gap that often goes overlooked in the existing literature.

2.2 Fuzzy-Logic Model

Fuzzy logic, conceived as a solution for managing systems characterized by ambiguity and vagueness, has been considerably integrated into the domain of pricing strategies, especially when these strategies incorporate variables like recommendations and demand availability [15]. Traditionally, forecasting in predictive modeling often leaned on deterministic models such as ARIMA [16], Exponential Smoothing State Space Model (ETS) [17], and the Prophet forecasting tool [18]. However, as market complexities intensified, so did the requirements for more adaptive tools. In this light, fuzzy logic highlighted its capabilities. [19] Were the first to propose fuzzy method to handle real dataset. After, a lot of studies emphasized fuzzy logic proficiency over classical models like ARIMA, ETS, and Prophet, particularly when faced with erratic and rapidly changing markets, and when coupled with other models, especially deep learning models ([20, 21]). For instance, while ETS and Prophet excel in capturing data patterns and seasonality, they occasionally stumble in dynamic pricing contexts, where fuzzy logic models maintain consistent performance due to their ability to process multi-dimensional inputs.

Furthermore, other methods, such as the Grey model and Theta model, while effective in certain forecasting scenarios, were found to be less efficient than fuzzy logic in others, as demonstrated by [22].

Regarding the forecasting task, fuzzy systems have demonstrated superiority over conventional models, leading to heightened prediction accuracy [23].

Yet, the sophistication of fuzzy logic also births its challenges. [24] Highlighted that the complexity making fuzzy logic adaptive can introduce computational demands and reduce the model's transparency. Despite its potential pitfalls, the intersection of fuzzy and traditional models offers a comprehensive perspective on pricing strategy, a notion we aim to further in our research.

3 Background

In this section, we provide a concise overview of the foundational principles and historical developments that inform the methodologies adopted in our study.

3.1 BiLSTM-Attention Model

LSTM and BiLSTM

Long Short-Term Memory (LSTM) was conceived by [25] as an innovative design within recurrent neural networks (RNN), equipped with a gradient-driven learning algorithm. Its primary intention was to navigate the challenge of sustaining long-term dependencies.

The conventional RNNs grapple with gradient diminishing and inflation issues. To address these, LSTM networks introduced the concept of "memory".

Refer to Fig. 1 for a contemporary depiction of LSTM cells [26].

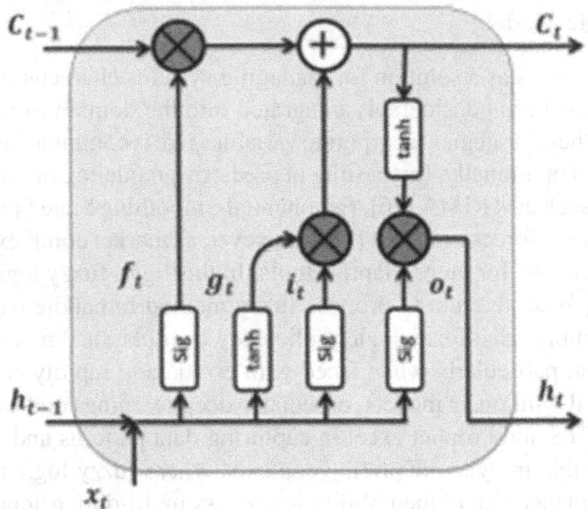

Fig. 1. Current Layout of LSTM

The LSTM units process three unique vectors. While the first vector, known as the input vector, is externally sourced and fed into the unit at time t, the remaining two emanate from the LSTM unit's previous state (t-1). They represent the hidden state and cell state.

The memory cell of the LSTM at a given time, t, is influenced by the external input as well as the brief memory from the last time step, marked by $x_{(t-1)}$ and $h_{(t-1)}$ respectively. By managing the flow of information, the LSTM gate navigates the challenge of diminishing gradients. Furthermore, BiLSTM integrates two single-directional LSTMs, each moving in a different direction: one forward and the other backward. This configuration ensures the network can access both current and past data.

Refer to Fig. 2 for visual depictions of the standard and bidirectional LSTM models.

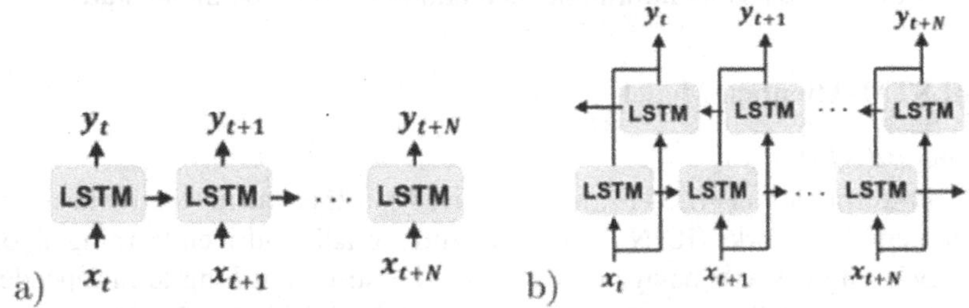

Fig. 2. a) Standard LSTM Model; b) BiLSTM Model.

The Attention Mechanism

The attention mechanism in deep learning is essentially a method allowing the model to zero in on pertinent features during its learning phase. It empowers the model to assign diverse weights to different input features based on their relevance.

The mechanism, in essence, computes attention scores for each input or time stamp, reflecting the significance of each feature. Subsequently, these inputs and their respective attention scores are merged to form a weighted input overview for the succeeding model layers.

A mathematical representation of the attention mechanism can be visualized as: Given a series of input vectors, symbolized as $X = [x_1, x_2, \ldots, x_n]$, and a target vector, represented as y, the attention sequence unfolds as:

- The input vectors are converted to a series of key and value vectors through adjustable parameters Wk and Wv.
- The target vector, y, undergoes transformation into a query vector, marked as Q, using adjustable parameters Wq.
- Attention scores, labelled as a, are determined through the scalar product of the query and key vectors.
- These scores are subsequently normalized using the softmax function, ensuring a total value of 1 for the scores.
- The attention weight, represented as w, is derived from the weighted average of value vectors using attention scores.
- The final output, z, is the combined weighted average of these value vectors.

This output is then routed through a connected layer, resulting in the ultimate prediction.

3.2 Fuzzy-Logic Model

The Fuzzy logic, offers a mathematical framework tailored to encapsulate the imprecise nature of human reasoning [15]. While classical or Boolean logic posits strict binary true-false evaluations, fuzzy logic permits intermediate truth-values, catering to systems infused with vagueness or ambiguity.

Fundamental Concepts and Operators in Fuzzy Logic

- Fuzzy Sets: Contrasting classic sets, which define clear membership boundaries, fuzzy sets, assign membership values ranging between 0 and 1 to depict gradient belongingness [15].
- Linguistic Variables: Such variables utilize descriptive terms as values rather than numerical ones. For instance, "Temperature" could be categorized as "cold", "warm", or "hot" [27].
- Membership Functions: They delineate the extent to which an item belongs to a particular fuzzy set [28].
- Fuzzy Logic Operators: Beyond the basic AND, OR, and NOT, fuzzy logic integrates operators that differ fundamentally from Boolean counterparts. The fuzzy AND, for instance, represents the intersection of two fuzzy sets through the minimal membership value [28].

Fuzzy Inference Systems
A Fuzzy Inference System (FIS) translates input values to determinable outputs. This encompasses:

- Fuzzification: This step transforms definitive input values into fuzzy ones using membership functions.
- Rule Evaluation: Rules are applied, such as "If Temperature is high AND Humidity is elevated, Comfort is minimal" [27].
- Aggregation: This stage amalgamates the outputs from varied rules.
- Defuzzification: Here, fuzzy output values revert to clear-cut values.

With the burgeoning advancements in artificial intelligence and machine learning, fuzzy logic steadfastly stands as a pivotal tool for grappling with ambiguity and uncertainty. It excels in mimicking human-like reasoning, proving essential for systems deeply intertwined with real-world nuances [15]. Specifically in the realm of decision-making, fuzzy logic offers a sophisticated framework to navigate the intricacies and uncertainties, rendering it invaluable for systems seeking refined discernment amidst ambiguous or imprecise information.

4 Material and Methods

In this section, we first present the proposed framework (Fig. 3):

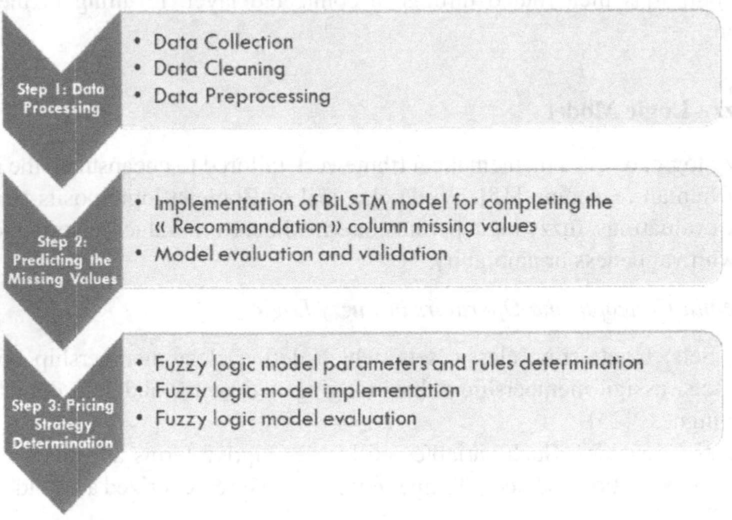

Fig. 3. Proposed Framework

4.1 Data Processing

For this study, we utilized a comprehensive dataset spanning four years (2019–2022) obtained from the Enterprise Resource Planning (ERP) system of a Moroccan car distributor, focusing specifically on their after-sales service. To ensure data integrity, we conducted cleansing procedures to remove duplicates, negative margins, blank cells, and outliers. After this process, the final dataset consisted of 50,095 rows, with 12 standardized columns.

In the preprocessing stage, we adopted various techniques to transform categorical variables into numerical values. This was accomplished using encoding methods. Additionally, to ensure uniformity and comparability, we normalized the data using the Min-Max method. By applying this technique, we transformed the data into a range of 0 to 1, employing the formula:

$$x_{min} = \frac{(x - x_{min})}{(x_{max} - x_{min})} \tag{1}$$

Subsequent to data preprocessing, the refined data was cast into 32 or 64-bit integers. It was then partitioned into two distinct subsets: a training set comprising 40,076 samples and a testing set with 10,019 samples. These data subsets served as the foundation for model training and performance evaluation.

4.2 Missing Values Prediction

For the effective prediction of missing values in sequential data, we deploy a multilayer BiLSTM architecture. This structure leverages BiLSTM's ability to recognize patterns and dependencies in both forward and backward time steps, thus providing comprehensive context to the model.

Below, Table 1 presents a detailed overview of the proposed BiLSTM model's architecture and its associated hyperparameters:

The implementation was executed in a Jupyter Notebook environment using the Python programming language, harnessing a BiLSTM-Attention architecture for in-depth analysis. The model was built using the Keras and TensorFlow libraries, offering a streamlined Python interface. Each layer in the architecture had a distinct role:

- The BiLSTM layers, defined by a specified "Number of layers", aimed to capture intricate patterns and dependencies in the data. Each layer used an activation function, "Relu" (Rectified Linear Unit), to transform the input signal further.
- "Batch normalization" was employed, a technique that applies normalization to a batch of data, enhancing training speed by enabling the use of a higher learning rate.
- Training parameters included "Batch size", indicating the number of samples used per update, and "Epochs", denoting the total number of training cycles. The "Adam" optimizer, an advanced variant of the gradient descent method, was used for optimizing the model. Adjustments to the model parameters were made using a predefined "Learning rate", determining the magnitude of changes during each gradient descent step towards minimizing the error function.

Table 1. Proposed BiLSTM model's architechture

Layer	Description
Input Layer	The data, after preprocessing, is fed into the model through this layer
BiLSTM Layer 1	- Number of units: 512 - Framework: Keras and TensorFlow - Return sequences: True - Batch Normalization: Stabilize and standardize activations
BiLSTM Layer 2	- Number of units: 256 - Framework: Keras and TensorFlow - Return sequences: True - Batch Normalization: Maintain consistent activations and smooth training
BiLSTM Layer 3	- Number of units: 128 - Framework: Keras and TensorFlow - Batch Normalization: Ensure activations remain within a desired range
Dense Layer	- Number of units: 64 - Activation function: ReLU - Framework: Keras and TensorFlow
Output Layer	- Number of units: 1 - Activation function: Linear - Framework: Keras and TensorFlow
Hyperparameters	
- Batch size: 100 - Epochs: 40 - Optimizer: Adam - Loss Function: Mean Squared Error (MSE) - Learning rate: 0.001	

4.3 Pricing Strategy Determination

In automotive after-sales service, optimizing pricing is influenced by various factors like demand, customer feedback, spare parts availability, and competitive rates. To better capture these intricacies, we introduce a tailored Fuzzy Logic Pricing Model. This model, adept at handling ambiguous data, translates these complexities into an algorithmic approach.

Here's the detailed algorithm:

ALGORITHM 1: Iterative Algorithm

```
BEGIN
  DEFINE Input Variables:
    Service Demand: Numerical value quantifying the need for specific services post car purchase.
    Customer Feedback: Numerical value indicating positive or negative feedback post service.
    Spare Parts Availability: Numerical value denoting the availability of parts needed for service.
    Competitive Service Pricing: Numerical value reflecting service pricing in comparison to competitors.
  FOR each Input Variable:
    MAP crisp value to fuzzy membership ('Poor', 'Average', 'Good') using membership functions.
  DEFINE Output Variable:
    Service Price: Suggested price for the after-sales service based on input factors.
  INITIATE Fuzzy Rule Base:
    COLLECTION of IF-THEN rules based on expert knowledge or historical service data.
  IMPLEMENT Fuzzy Inference System:
    Fuzzification:
      CONVERT crisp inputs to fuzzy values.
    Rule Evaluation:
      DETERMINE applicable rules and compute rule truth degree.
    Aggregation:
      COMBINE rule outcomes to produce comprehensive fuzzy output.
    Defuzzification:
      DERIVE single crisp output (recommended service price) from fuzzy output.
  IMPLEMENT Sensitivity Analysis Mechanism:
    FOR each Input Variable:
      ADJUST the variable's value.
      OBSERVE change in recommended service price.
      DETERMINE sensitivity based on service price change.
END
```

The Fuzzy Logic Pricing Algorithm for Automotive After-Sales Service offers a structured approach to determine optimal service pricing. Leveraging fuzzy logic, the algorithm considers ambiguous and qualitative factors, such as customer feedback and competitive rates. By converting these inputs into linguistic variables like 'Poor', 'Average', or 'Good', it can generate a pricing recommendation that is more aligned with real-world nuances. The inclusion of a sensitivity analysis further adds robustness, allowing stakeholders to understand how variations in individual factors impact the final price. In essence, this algorithm provides a holistic and adaptive method to derive pricing strategies in the automotive after-sales domain.

5 Results

The analyses were conducted on a Jupyter Notebook environment running on a system powered by an Intel(R) CORE(TM) i5 processor with 8 GB RAM. This setup also featured an Intel UHD Graphics GPU and operated on Windows 10 Pro 64-bit. With this computational framework in place, we now delve into the results obtained from our study in the subsequent section.

To properly understand the performance and accuracy of our models in the context of automotive after-sales service pricing, we employed two distinct evaluation metrics. Both our fuzzy logic and BiLSTM-attention models were assessed using these criteria. The primary metric adopted was the "Mean Squared Error" (MSE), defined as:

$$MSE = \frac{\sum(yi - yp)^2}{n} \tag{2}$$

The MSE provides a measure of the average squared difference between predicted and actual values, indicating the model's accuracy.

Another crucial metric used for assessment was R2, which gauges the proportion of the variance in the dependent variable predicted by the independent variables. It is given by:

$$R^2 = 1 - \frac{\sum(yi - yp)^2)/n}{\sum(\hat{y}i - yp)^2)/n} \tag{3}$$

In these equations, y_i represents the predicted value, y_p stands for the actual or true value, $\hat{y}_i$ is the mean of observed data, and n is the total number of observations or entries. The following sections provide a detailed breakdown of the results obtained using these metrics across our models.

5.1 BiLSTM-Attention Model

The BiLSTM-attention proposed model showed good performances. In fact, Fig. 4 elucidates its journey of adaptation throughout the training phase. Initially, there is a noticeable discrepancy in training error, signaling the model's fledgling efforts to capture the intricacies of the dataset. However, as the training advances, this error substantially reduces, highlighting the model's improving proficiency. Concurrently, the validation error, starting at a higher value, declines swiftly before reaching a steady state, indicating an enhanced ability of the model to generalize on unfamiliar data. The ultimate convergence of both training and validation curves at a similar error rate is a testament to the model's balanced learning, successfully steering clear of pitfalls like overfitting or underfitting.

In order to further evaluate the efficacy of our BiLSTM-attention model, we compared its performance against BiLSTM and LSTM models. The results of this comparative analysis are presented in Table 2 as follow:

The BiLSTM-attention model outperforms the others with the lowest MSE of 0.041 and a high R2 score of 0.87, indicating its predictions are closest to the actual values and it can explain 87% of the variance. The BiLSTM follows closely with an MSE of

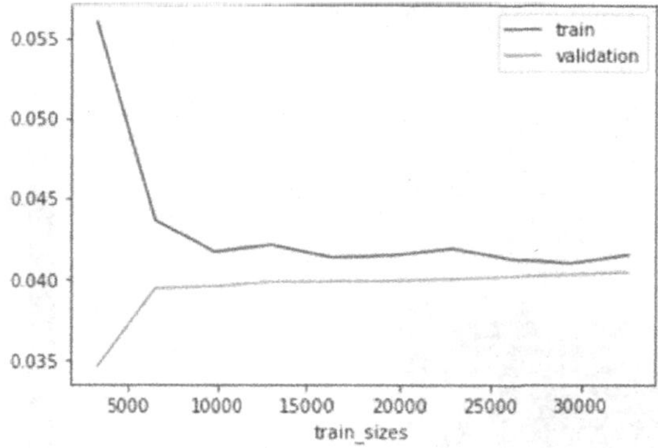

Fig. 4. Line plot of train and validation loss from the BiLSTM –attention model

Table 2. Criteria metrics results depending on model

	MSE	R2
BiLSTM-attention	0.041	0.87
BiLSTM	0.045	0.85
LSTM	0.053	0.78

0.045 and R2 of 0.85. The LSTM model, though competent, shows slightly inferior performance with an MSE of 0.053 and R2 of 0.78, indicating a higher deviation in its predictions and lesser explanatory power.

5.2 Fuzzy Model

To understand the performance of our model, we conducted a sensitivity analysis, visualizing the intricate interplay between "Demand," "Recommendation Sentiment," and the resulting "Output Price." The 2D plot presented in Fig. 5 sheds light on various facets:

At the low end, depicted by the lower-left quadrant, both "Demand" and "Recommendation Sentiment" are minimal, leading to the lowest "Output Price," evident from the deep purple hue. Products with neither demand nor favorable feedback should thus be priced modestly, perhaps as an incentive for potential buyers.

Interestingly, the model showcases a heightened sensitivity to "Recommendation Sentiment" over "Demand." Even with average demand, a surge in positive sentiment can significantly elevate the proposed price, emphasizing the influence of commendable reviews.

While traversing regions of increased demand, the price remains fairly consistent until a rise in sentiment supports it. This highlights that mere demand isn't enough for a pronounced price hike; positive feedback is pivotal.

The pinnacle of pricing, symbolized by the top-right corner's rich green shade, is reached when a product is both highly sought-after and garners stellar recommendations.

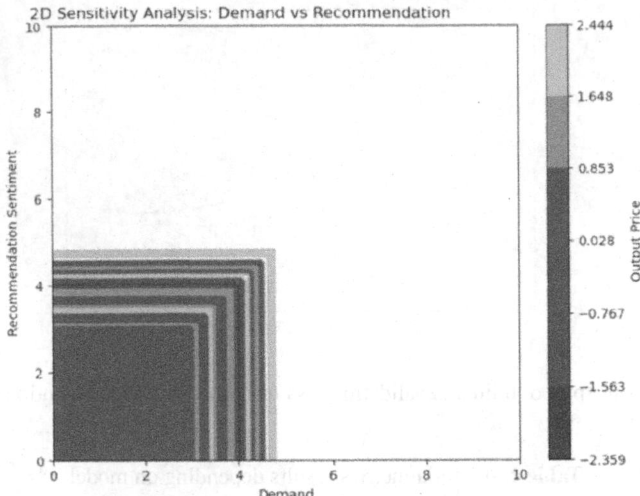

Fig. 5. Sensitivity Analysis

However, this zone also reflects diminishing returns, with densely packed contour lines indicating that further boosts in demand or sentiment yield minimal price increments.

Overall, while demand is undeniably important, the analysis accentuates the pivotal role of positive sentiment in driving prices. Businesses should thus prioritize not just amplifying demand, but also securing positive endorsements to fine-tune their pricing strategies.

To further evaluate the model's performance, Fig. 6 & Fig. 7 provide insightful visualizations.

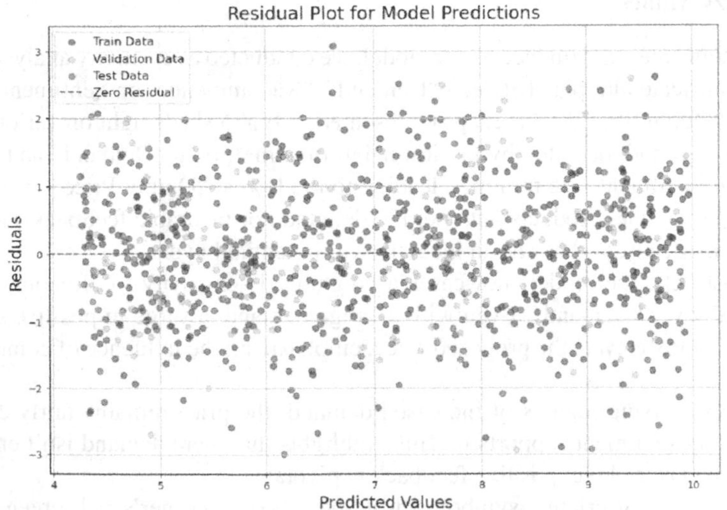

Fig. 6. Residual Plot

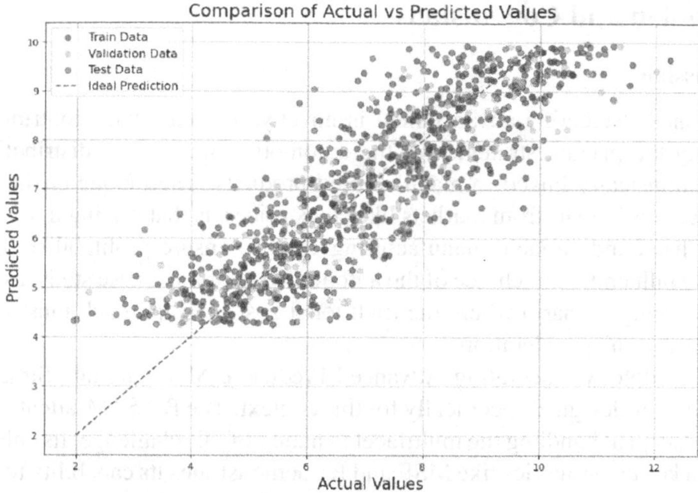

Fig. 7. Actual vs Predicted values

The Residual Plot highlights a random scattering of residuals around the zero line, suggesting consistent, unbiased predictions across Train, Validation, and Test datasets. Moreover, the homoscedastic nature of the residuals indicates a steady variance across predicted values. In the Actual vs Predicted Values Plot, the proximity of points to the red line of perfect prediction underscores the model's accuracy. The dense clustering in the mid-range signifies precise predictions for a majority of standard values, while the minimal spread at higher values confirms its robustness across various data points. Both plots collectively highlight the model's commendable accuracy and reliability.

Regarding the evaluation metrics, our model exhibited exceptional performance as demonstrated by the following results:

- Mean Squared Error (MSE): The model achieved an MSE of 0.0730, which indicates a relatively low average squared difference between the actual and predicted values. A lower MSE implies better accuracy, suggesting that our model's predictions are closely aligned with the actual outcomes.
- R-squared (R2): With an R2 value of 0.91, our model explains approximately 91% of the variance in the dependent variable. This high R2 value underscores the model's ability to capture the underlying patterns in the data. An R2 value close to 1 indicates that the model is well-fitted to the observed data points.

These metrics highlight the model's robustness and precision in predictions, although there is always room for improvement, especially in reducing the MSE further.

6 Discussion and Conclusions

6.1 Discussion

A cornerstone of scientific advancement is novelty, and our study contributes to this by exploring the pricing strategies in the automotive spare parts distribution sector, a domain intrinsically linked to manufactured products. The unique challenge in this sector, which sets it apart from markets like stock prices, is that pricing must adequately cover both fixed and variable manufacturing costs to ensure profitability. This aspect significantly influences our choice of the automotive sector as a case study, as it presents a complex pricing scenario where the traditional cost-plus approach must be nuanced with market-driven considerations.

Our methodology, integrating Advanced Predictive Modeling and Strategic Decision Modeling, is designed specifically for this context. The BiLSTM-attention model is particularly adept at handling the multifaceted nature of this challenge. Its enhanced performance in key error metrics like MSE and R2 demonstrates its capability to accurately forecast prices that not only align with manufacturing costs but also adapt to dynamic market conditions. This is a significant departure from simpler predictive models and highlights the applicability of our approach in a sector where the pricing strategy is a delicate balance between covering manufacturing costs and remaining competitive in the market.

Additionally, the incorporation of Fuzzy Logic in our model addresses the intricacies of pricing in the automotive distribution sector. Unlike traditional models, which may primarily focus on manufacturing costs, Fuzzy Logic allows us to factor in a broader range of variables such as customer demand trends, stock conditions, and distribution logistics. This holistic approach is crucial in a sector where prices are not solely determined by production costs but are also influenced by market dynamics and consumer behavior.

However, we acknowledge the limitations of our approach, particularly in the context of unexpected market fluctuations that are common in the automotive sector. While the BiLSTM-attention model provides a robust framework for price prediction, it may not fully account for abrupt market changes. Similarly, the broad scope of Fuzzy Logic, while advantageous in capturing a range of factors, may require further refinement to address specific challenges that arise from the complex nature of pricing manufactured products like automotive spare parts.

6.2 Conclusions

In the vast ocean of data-driven decision-making methodologies, our approach emerges as a pioneering beacon, particularly in the realm of automotive spare parts distribution, a sector grappling with the unique challenges of pricing manufactured products. Our study illustrates the profound impact of blending state-of-the-art predictive and strategic models to address the intricacies of such a complex sector. Through rigorous testing and validation, both through error metrics and real-world API application, we've unveiled a framework that not only raises the bar in forecasting accuracy but also promises transformative implications for businesses in sectors where pricing needs to consider both manufacturing costs and market dynamics.

The potential of our approach to recalibrate and redefine pricing strategies in the automotive distribution sector makes it an indispensable asset in today's data-centric business landscape. As we move forward, we envision further refining and expanding this hybrid model. Our focus will be on incorporating real-time data feeds to better capture the rapidly changing market conditions specific to automotive distribution, integrating more granular parameters that reflect the complexities of this sector, and exploring adaptive learning mechanisms to continuously enhance its accuracy and applicability.

Future research avenues also include investigating the framework's applicability across different industry verticals, particularly those dealing with the pricing of manufactured goods, and scaling the framework to handle even larger datasets. This continuous evolution will undoubtedly solidify its foothold as a paramount tool in modern analytics, especially in sectors like automotive spare parts distribution where the balance between manufacturing costs, market trends, and customer demand is vital for successful pricing strategies.

References

1. Kreuzer, T., Röglinger, M., Rupprecht, L.: Customer-centric prioritization of process improvement projects. Decis. Support Syst. **133**, 113286 (2020)
2. Owramipur, F., Eskandarian, P., Mozneb, F.S.: Football result prediction with Bayesian network in Spanish League-Barcelona team. Int. J. Comput. Theory Eng. **5**(5), 812 (2013)
3. Yusof, Y., Mustaffa, Z.: Dengue outbreak prediction: a least squares support vector machines approach. Int. J. Comput. Theory Eng. **3**(4), 489 (2011)
4. Hariri, R.H., Fredericks, E.M., Bowers, K.M.: Uncertainty in big data analytics: survey, opportunities, and challenges. J. Big Data **6**(1), 1–16 (2019)
5. Abayomi-Alli, O., Misra, S., Abayomi-Alli, A.: A deep learning method for automatic SMS spam classification: performance of learning algorithms on indigenous dataset. Concurrency Comput. Pract. Exper. **34**(17), e6989 (2022)
6. Lin, C.T., Chiu, H., Tseng, Y.H.: Agility evaluation using fuzzy logic. Int. J. Prod. Econ. **101**(2), 353–368 (2006)
7. Huo, B., Zhao, X., Zhou, H.: The effects of competitive environment on supply chain information sharing and performance: an empirical study in China. Prod. Oper. Manag. **23**(4), 552–569 (2014)
8. Dudek, G.: Short-term load forecasting using random forests. In: Filev, D., Jabłkowski, J., Kacprzyk, J., Krawczak, M., Popchev, I., Rutkowski, L., Sgurev, V., Sotirova, E., Szynkarczyk, P., Zadrozny, S. (eds.) Intelligent Systems'2014. AISC, vol. 323, pp. 821–828. Springer, Cham (2015). https://doi.org/10.1007/978-3-319-11310-4_71
9. Amellal, A., Amellal, I., Seghiouer, H., Ech-Charrat, M.R.: Improving lead time forecasting and anomaly detection for automotive spare parts with a combined CNN-LSTM approach. **2**(16), 265–278 (2023)
10. Lu, W., Li, J., Li, Y., Sun, A., Wang, J.: A CNN-LSTM-based Model to Forecast Stock Prices.," pp. 1–10 (2020)
11. Vaswani, A., et al.: Attention is all you need, pp. 5998–6008 (2017)
12. Wang, S., Wang, X., Wang, S., Wang, D.: Bi-directional long short-term memory method based on attention mechanism and rolling update for short-term load forecasting. **109**, 470–479 (2019)
13. Gomez, W., Wang, F.K., Amogne, Z.E.: Electricity load and price forecasting using a hybrid method based bidirectional long short-term memory with attention mechanism model (2023)

14. Zadeh, L.A.: Fuzzy logic—a personal perspective. Fuzzy Sets Systems, **281**, 4–20 (2015)
15. Makridakis, S., Hibon, M.: ARMA models and the Box-Jenkins methodology. J. Forecast. **16**(3), 147–163 (1997)
16. Hyndman, R.J., Khandakar, Y.: Automatic time series forecasting: the forecast package for R. J. Stat. Softw. **27**, 1–22 (2008)
17. Taylor, S.J., Lethman, B.: Forecasting at scale. Am. Stat. **72**(1), 37–45 (2018)
18. Wang, L.X., Mendel, J.M.: Generating fuzzy rules by learning from examples. IEEE Trans. Syst. Man Cybernet. **22**, 1414–1427 (1992)
19. Iqbal, S., Zhang, C., Arif, M., Wang, Y., Dicu, A.M.: A comparative study of fuzzy logic regression and ARIMA models for prediction of gram production. In: Balas, V.E., Jain, L.C., Balas, M.M., Shahbazova, S.N. (eds.) SOFA 2018. AISC, vol. 1222, pp. 289–299. Springer, Cham (2021). https://doi.org/10.1007/978-3-030-52190-5_21
20. Imani, M.: Fuzzy-based weighting long short-term memory network for demand forecasting. J. Supercomput. **79**(1), 435–460 (2023)
21. Sfetsos, A.: A comparison of various forecasting techniques applied to mean hourly wind speed time series. Renew. Energy **21**(1), 23–35 (2000)
22. Kambalimath, S., Deka, P.C.: A basic review of fuzzy logic applications in hydrology and water resources. Appl. Water Sci. **10**(8), 1–14 (2020)
23. Gandhmal, D.P., Kumar, K.: Systematic analysis and review of stock market prediction techniques. Comput. Sci. Rev. **34**, 100190 (2019)
24. Hochreiter, S., Schmidhuber, J.: Long short-term memory. Neural Comput. **9**(8), 1735–1780 (1997)
25. Smagulova, K., James, A.P.: A survey on LSTM memristive neural network architectures and applications. Eur. Phys. J. Spec. Top. **228**(10), 2313–2324 (2019)
26. Mamdani, E.H.: Application of fuzzy logic to approximate reasoning using linguistic synthesis. IEEE Trans. Comput. **12**, 1182–1191 (1977)
27. Klir, V., Yuan, B.: Fuzzy sets and fuzzy logic. Prentice hall New Jersey (1995)
28. Amellal, I., Amellal, A., Seghiouer, H., Ech-Charrat, M.: An integrated approach for modern supply chain management: utilizing advanced machine learning models for sentiment analysis, demand forecasting, and probabilistic price prediction. Decis. Sci. Lett. **13**(1), 237–248 (2024)

Full-Length Hardness Prediction in Wire Rod Manufacturing Using Semantic Segmentation of Thermal Images

Seok-Kyu Pyo[1], Sung-Jun Hur[1], Dong-Hee Lee[1(✉)], Sang-Hyeon Lee[2], Sung-Jun Lim[2], Jong-Eun Lee[2], and Hong-Kil Moon[2]

[1] Department of Industrial Engineering, Sungkyunkwan University, Suwon-si, Republic of Korea
dhee@skku.edu

[2] Electric Furnace Process Research Team, Hyundai Steel R&D Center, Dangjin-si, Republic of Korea

Abstract. As an essential steel product, wire rods have specific requirements regarding their physical properties. Especially for wire rods for automotive springs, it is important to ensure consistent hardness throughout the product. Because traditional hardness testing methods are destructive and sample-based, they have the potential to overlook the non-uniformity of wire rod hardness. This paper presents the application of a convolutional neural network (CNN) to thermal imaging to address these issues. The model segments the thermal image of a wire rod after cooling, separating the temperature of the wire rod and the background on a pixel-by-pixel basis. This temperature data is used to calculate the cooling rate and helps to predict the hardness of the wire rod along its entire length. Experimental results show that the U-Net-based model outperforms a simple FCN model in the segmentation task. This approach provides a more comprehensive quality inspection of wire rod, bringing both economic and quality benefits to the steel industry.

Keywords: wire rod · semantic segmentation · hardness prediction · thermal image

1 Introduction

A wire rod is a semi-finished steel product that must meet specific requirements in terms of its physical properties. Produced through steelmaking, continuous casting, billet rolling, and wire rod rolling processes, wire rods are further processed into finished products such as wires, springs, bolts, and nuts. Noncompliance with the specifications for the physical properties of the wire rod may result in processing failures or a decline in the quality of the finished product. Its representative physical properties are fatigue strength, tensile strength, yield strength, and hardness [1]. Hardness is a major factor, particularly in automotive suspension systems. When the hardness does not satisfy the specifications, the fatigue strength and ductility typically decrease [2–4]. A reduction in ductility can result in damage during spring drawing, and a decrease in fatigue strength negatively affects the durability and safety of the suspension system.

S.-H. Sheu (Ed.): ICIEA-EU 2024, LNBIP 507, pp. 189–199, 2024.
https://doi.org/10.1007/978-3-031-58113-7_16

In general, the hardness of a wire rod is measured on a single section of a wire rod out of 5 to 10 coils produced from the same semi-finished product (bloom) for wire rod hardness quality inspection. However, wire rods can be hundreds to thousands of meters long, and the hardness varies greatly from section to section. Therefore, a product that passes the sampling test may have hardness defects in other sections. Since hardness testing is a destructive test, it is not possible to perform an total inspection, but it is possible to overcome this problem by using process history data to predict full-length hardness quality.

Among the wire rod production processes, the wire rod rolling process plays the most crucial role in determining the hardness of the wire rod because it is the final process of wire rod manufacturing and the point of the final phase transformation. In particular, the cooling process, which is a subprocess of the wire rod rolling process, has the greatest impact on hardness because phase transformation occurs during this process. The cooling rate of the wire rod during phase transformation is a critical factor affecting the hardness. The cooling rate is typically calculated as the difference between the temperature of the wire rod measured before and after the cooling process. In addition to the cooling rate, the alloy composition set in the steelmaking process also affects the hardness of the wire rod [5].

In order to calculate the cooling rate of a wire rod, it is necessary to measure the pre-cooling and post-cooling temperatures of the material. The pre-cooling temperature is easy to measure because the material is quickly processed into a straight line and the temperature is constant until the cooling process begins. However, the cooling process takes place after the part has been processed into a coil in the laying head. Unlike billets, bars, or flat steel plates, which are simple shapes, wire rods are not uniform and have complex geometric curves, making it difficult to measure the post-cooling temperature of the material. Even if a thermograph is obtained using a thermal imaging instrument, the thermal image does not contain any information to distinguish between the material and non-material parts. Since the human eye can only distinguish between materials based on their appearance in the thermal image, engineers have traditionally manually collected some of the material temperature from the thermal image.

In this study, we propose to utilize semantic segmentation methods to segregate and collect material-specific post-cooling temperatures from thermal images and predict the full-length hardness of wire rods. The thermal images are divided into 128×128 pixel image patches and manually annotated to generate training data, and a convolutional neural network (CNN) is used to train a segmentation model to determine the material. The cooling rate is calculated from the post-cooling temperature extracted through the prediction model and the pre-cooling temperature and cooling time from the process history. Using the process factors including the cooling rate, the hardness prediction model developed in previous studies is used to predict the full-length hardness. This method is expected to complement the limitations of quality inspection that relies on random sampling.

The remainder of this paper is organized as follows: Sect. 2 introduces the wire rod manufacturing process and reviews the relevant existing studies. Section 3 describes the proposed approach for developing the wire rod semantic segmentation model. Section 4 summarizes the proposed approach and discusses future research directions.

2 Background

2.1 Wire Rod Manufacturing Process

A flow diagram of the wire rod manufacturing process is shown in Fig. 1. There are four main processes for manufacturing wire rods: steelmaking, continuous casting, billet rolling, and wire rod rolling. During steelmaking, scrap iron is transformed into molten iron using an electric arc furnace and then further heated in a furnace. To adjust the temperature and components of the molten iron, it was passed through a ladle furnace and a Rheinstahl–Heraus furnace (RH). The molten steel is solidified through a continuous casting process and turned into semi-finished products, called billets, through the billet rolling process.

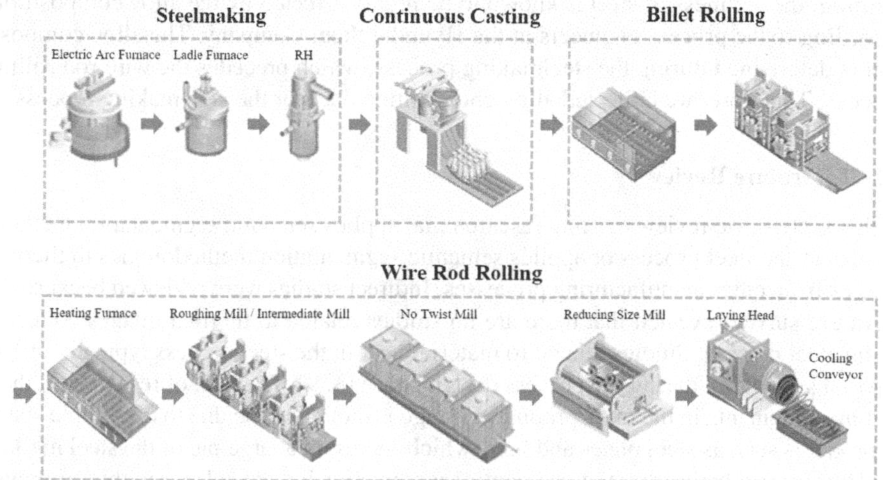

Fig. 1. Wire rod manufacturing process.

In the wire rod rolling process, a heating furnace heats the billet from room temperature to over 900°C to facilitate hot rolling. The heating furnace determines the overall temperature of the billets during the wire rod rolling process because there are no further heating processes. The heated billet undergoes a pressurized water spray process known as a descaler to remove impurities such as an iron oxide layer from the surface of the billet before moving to the continuous rolling mills. Continuous rolling mills transform the billet into a straight wire rod of 5–25 mm in diameter. In roughing and intermediate mills, a rectangular billet is deformed into a circular or oval cross-section and then passed through a no-twist mill, reducing the size of the mill to be rolled into a shape with the target diameter. The straight wire rod is uniformly cooled to the target temperature using water-cooling boxes and laid into circular loops on the cooling conveyor through the laying head facility. Cooling conveyors cool the overlapping wire rods in the form of circular loops with forced air using a series of fans to obtain the desired microstructure and mechanical properties.

The wire rod rolling process, which is the final stage of wire rod production, has a significant impact on the hardness of wire rods. Factors such as the heating furnace temperature and the percentage reduction in continuous rolling mills affect the microstructure and physical properties of steel [6]. In the cooling conveyor, the final phase transformation of the wire rod occurs, which determines the physical properties. The cooling rate is the factor that affects hardness the most of all process factors [7]. The cooling rate depends on whether the conveyor cover is open or closed, the blower nozzle and blowing volume, the spacing of the wire rod rings, etc. [8]. Hence, operational data including temperature and cooling rate data of wire rods are collected in this wire rod rolling process. The temperature before cooling is measured at the laying head, and the temperature after cooling is collected in the form of thermal images from a line scanner installed at the rear end of the cooling conveyor. Operational data were collected from the production line of Hyundai Steel Company, a major steel company in Korea. In addition, the hardness of steel is known to be highly affected by the alloy composition, according to the process engineers at the Hyundai Steel Company. The alloy composition is determined during the steelmaking process, which precedes the wire rod rolling process. Therefore, we collected alloy composition data for the steelmaking process.

2.2 Literature Review

In this section, we review existing research that applies semantic segmentation methodologies to the steel process or applies semantic segmentation methodologies to thermal images from other manufacturing processes. Indirect studies were reviewed because the literature survey revealed that there are no studies related to thermal images collected in the steel process. Studies related to material heat in the steel process typically utilize thermal simulation during the process design phase [8, 9]. The lack of research on thermal measurements in the mass production stage is thought to be due to the simple shape of products such as steel plates and bars, which represent a large pie of the steel market.

Before introducing the study, semantic segmentation is a crucial area within computer vision that involves classifying each pixel in an image to a particular class. Unlike image classification, where the task is to classify an entire image into a single category, or object detection, where specific objects within an image are detected and bounded, semantic segmentation aims to provide a detailed pixel-wise labeling of an image. Fully convolutional network (FCN) is the standard model in the field, and models such as U-Net and DeepLabV3 + are widely used. In particular, U-Net is the most referenced model in semantic segmentation with over 30,000 cites of its original paper. The name "U-Net" comes from its u-shaped architecture as a result of a symmetric encoder-decoder. The original model was created to segment cells in biomedical images [10].

In the steel sector, semantic segmentation has been used for steel scrap classification and steel microstructure classification. Daigo et al. (2023) apply semantic segmentation methods to images of collected steel scrap to classify the type of steel scrap without directly measuring the thickness and diameter of the steel [11]. Different scraps have different scales and textures, which can lead to poor performance when predicted by FCN. Therefore, the Pyramid Scene Parsing Network (PSPNet), which introduces a pyramid pooling module to integrate multiple feature resolutions, is applied. Laub et al. (2022), Thomas et al. (2020), and Xie et al. (2023) applied semantic segmentation methods for steel microstructure classification [12–14]. Both Laub et al. (2022) and Thomas et al. (2020) used U-Net. Laub et al. (2022) segmented the image to be analyzed into individual patches for training and prediction, and used image enhancement techniques such as rotation and flip. Xie et al. (2023) propose an improved fully convolutional network ASPP-FCN for automatic identification of multiphase microstructures in steel, and carry out comparative experiments with different networks, including FCN, DeepLab v3 +, Unet, Enet and PSPnet.

Examples of applications of semantic segmentation methodologies to thermal images collected from manufacturing processes in other industries include aluminum temperature measurement and carbon fiber non-destructive testing. Xu et al. (2028) distinguish between electrolyte and floaters to accurately measure the temperature of electrolyte in infrared thermography images of electrolyzed aluminum in an aluminum plant. The problem is that the shape of the electrolyte is not consistent and the distribution and covering of the electrolyte is irregular, which is solved by utilizing the DeepLab network, an improved model of FCN. The same researchers, Lema et al. (2023) and Pedrayes et al. (2022), apply semantic segmentation to thermography, a non-destructive methodology for detecting subsurface defects in carbon fibers. They apply post-processing methodologies such as Principal Components Thermography (PCT) to U-Net and DeepLabV3 + models to improve performance.

The literature review showed that semantic segmentation is applied to thermal images when the material shape is inconsistent or when the temperature directly affects the product quality, and furthermore, when temperature measurement is a quality inspection. Various models were used, but FCN's improved models U-Net and DeepLabV3 + were the most frequently used. Preprocessing techniques such as image patch segmentation and enhancement, and postprocessing techniques such as PCT were used. The application of semantic segmentation to thermal images is appropriate because wire rods are geometrically complex, their placement is not uniform, and the cooling rate directly affects product quality. It was also decided to utilize the U-Net model and preprocessing techniques that have been validated in other studies.

3 Development of a Wire Rod Semantic Segmentation Model and Full-Length Hardness Prediction

In this section, the details of the proposed approach for developing a hardness predic-tion model predicting full-length hardness are explained. The proposed approach consists of four steps. The five steps were conducted in close collaboration with Hyundai Steel's Electric Furnace Process Research Team. The team is in charge of quality control for electric furnace-based steel production processes, including steelmaking, rolling, surface treatment, and control and measurement technologies. Each step is explained in detail below.

3.1 Step 1. Collecting the Wire Rod Thermal Images

Thermal images of wire rods were collected in real time during the wire rod rolling process at Hyundai Steel. A thermal line scanner installed at the back of the cooling conveyor automatically collects temperature data in a two-dimensional array. Five different images of wire rods with different diameter specifications (12 mm to 18 mm) were collected. Each product has a size of 1000 points horizontally and thousands to tens of thousands of points vertically, depending on the product size. From the data, a two-dimensional array of 512 points wide temperature data was taken where the wire rod was present in the data.

3.2 Step 2. Image Preprocessing and Annotation

To train an efficient CNN model with the limited data of 5 thermal images, we need to increase the number of data by dividing the image into patches. Wire rods have a symmetrical shape about the center of their width. Therefore, the images were divided into patches of 128×128 pixels to create four images in the horizontal plane.

Since the temperature data is in the form of a two-dimensional array, we need to apply a color filter to convert it to a colored three-dimensional array. We applied a "BWR" filter because it is intuitive to the human eye to color high temperatures red and low temperatures blue. The "BWR" filter is applied based on the minimum and maximum values of the temperature. The color of the image varies depending on whether this filter is applied to the image before segmentation or to each image patch after segmentation.

Since infrared thermometers measure thermal radiation in the form of infrared light, when measuring a hot object, the radiant heat emitted from the object to the atmosphere at the edges of the object appears as a halo in the thermal image. In Fig. 2, the white band around the edge of the material is an example of this phenomenon. This white area is ambiguous as to whether it belongs to the material or to the space around the material. Since the goal of the semantic segmentation model is to extract the temperature of the wire rod only, it needs to learn that the boundary is not the wire rod for accurate annotation and classification. Figure 3 shows the color filter applied to the entire image before segmentation and to each patch after segmentation. The latter is used because the filter is applied to each patch and the distinction between wire rod, boundary, and background is clearer.

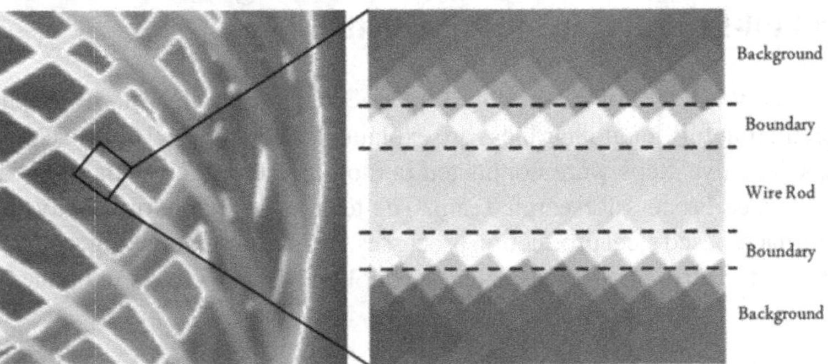

Fig. 2. Wire rod thermal image and components.

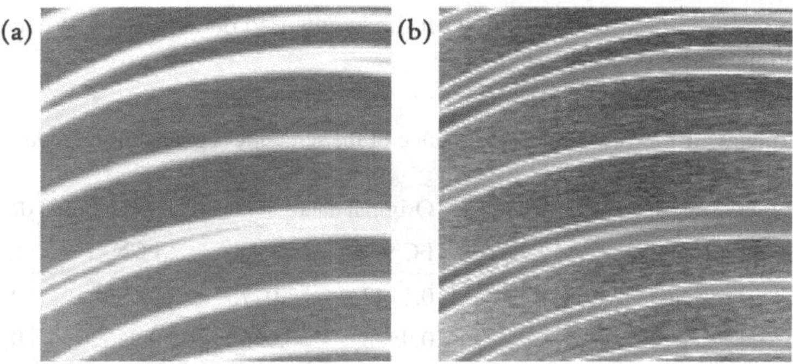

Fig. 3. Image patches depending on how the color filter is applied (a) Appling a single image color filter (b) Appling individual patch color filters.

The five wire rod thermograms were segmented into 1292 patches, which were manually annotated to reflect the visual determination of the wire rod. Figure 4 depicts the annotation method. Labeling of 340 patches was performed by three researchers. The patches were labeled into two classes: 'wire rod' to represent wire rod parts and 'background' to include non-wire rod parts and boundaries. The 340 patches labeled as wire rod or not were augmented 8x with a combination of rotation and flipping, resulting in a total of 2720 patches.

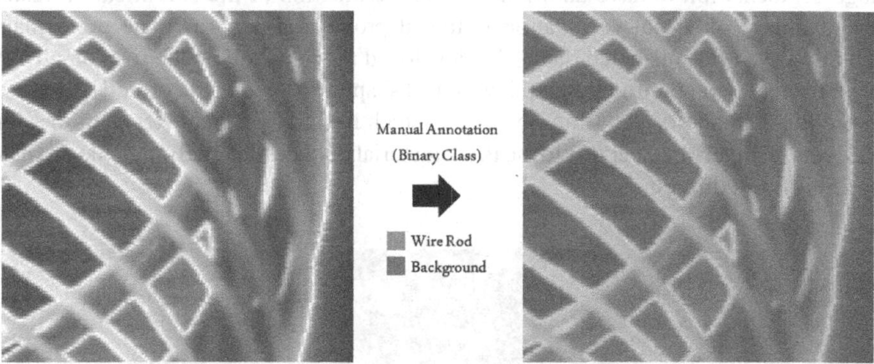

Fig. 4. Wire rod thermal image manual annotation.

3.3 Step 3. Training the Semantic Segmentation Model

A U-Net model based on a fully convolutional network (FCN), a CNN methodology, was trained on the image patch dataset preprocessed in Step 2. A simple FCN model was used as a comparison. The train: validation: test ratio was set to 7: 1: 2. Table 1 shows the classification performance of the models. Both the FCN and U-Net models improved their performance after data augmentation. On the same dataset, the U-Net model outperformed the FCN model on all but the test loss value. In the performance evaluation of the U-Net model trained on the augmented dataset, the test accuracy was 0.9636 and the test F1-score was 0.9489.

Table 1. Comparing the performance of the semantic segmentation models.

		Original data		Augmented data	
		FCN	U-Net	FCN	U-Net
Train	Binary cross entropy	0.1147	0.0776	0.0808	0.0697
	Accuracy	0.9508	0.9668	0.9662	0.9696
	F1-score	0.9326	0.9532	0.9509	0.9559
Validation	Binary cross entropy	0.1652	0.1454	0.0822	0.0763
	Accuracy	0.9320	0.9483	0.9653	0.9674
	F1-score	0.9134	0.9318	0.9496	0.9528
Test	Binary cross entropy	0.1557	0.1258	0.0956	0.1004
	Accuracy	0.9422	0.9566	0.9613	0.9636
	F1-score	0.9105	0.9303	0.9455	0.9489

While quantitative performance evaluation is an objective metric, it is essential for image segmentation to have additional visual verification of the classification results. The advantage of visual verification is that it provides an approximate performance evaluation even for unlabeled data. The developed model made predictions on all 1292 image patches and confirmed that the wire rod shape was predicted correctly. However, the predicted image of the 18mm product, which has a larger diameter than the other sizes, was relatively unclear due to the thick material thickness. A portion of the predicted image is depicted in Fig. 5.

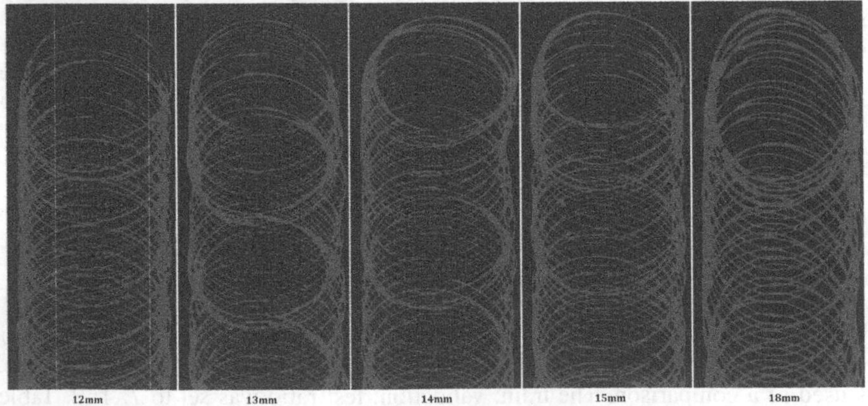

Fig. 5. Wire rod prediction images.

3.4 Step 4. Predicting the Wire Rod Hardness

Based on the coordinates of the pixels corresponding to the wire rods identified in Step 3, it was possible to extract the post-cooling temperature of the wire rods and calculate the cooling rate. The hardness prediction model developed in the previous study predicts the hardness of spring steel wire rods based on the cooling rate, pre-cooling temperature, and alloy composition. The prediction model can be used to predict the hardness of every pixel in the thermal image.

A graphical tool has been developed to communicate the full-length hardness quality of wire rod to field engineers. At each pixel, the wire rod hardness is colored green if it is in the normal range, yellow if it needs attention, and red if it is out of specification. Engineers can look at the image to determine how many rings to trim from each end of the wire rod, and whether to rework the product.

4 Conclusion and Future Research

This study represents a comprehensive approach to enhancing the quality inspection of wire rods, particularly focusing on the hardness aspect, by utilizing modern semantic segmentation methods. The central issue addressed was the inconsistency of the conventional hardness inspection method, which is based on a sampling approach, possibly overlooking defects in non-sampled sections. The use of thermal imaging, combined with the power of CNNs, especially the U-Net model, has shown potential in overcoming these limitations.

Several noteworthy achievements and observations were drawn from this research. The application of semantic segmentation in segregating the material from the background in thermal images of steel industry is innovative. Notably, the U-Net model outperformed the traditional FCN model in most aspects, particularly when trained on augmented data. The significant role of preprocessing, particularly the method of applying the "BWR" color filter, is highlighted. This preprocessing allowed for clearer distinction between the wire rod, its boundary, and the surrounding background. Beyond a mere academic exercise, this approach has direct implications in real-world scenarios. The collaboration with Hyundai Steel's Electric Furnace Process Research Team ensures that the proposed methods are grounded in real industrial requirements and can be practically implemented. The development of a visual representation, coloring wire rod hardness on a spectrum from green (normal) to red (out of specification), provides engineers with an intuitive tool to quickly assess product quality and make necessary adjustments.

However, as with all technological advancements, this study also hints at areas of future research, especially regarding the challenges posed by wire rods of differing diameters, such as the 18mm product, which showed some prediction discrepancies. In the case of 5.5 mm and 9 mm wire rods, the material is thinner and cools faster, which makes it difficult to clearly distinguish the wire rod parts with the naked eye in the thermal image. In these cases, the predictions made by the developed model are not accurate. For products with extreme specifications, we plan to conduct research on data augmentation using morphology and diffusion models for post-processing of predictive images.

In summary, by merging thermal imaging with state-of-the-art machine learning techniques, this research paves the way for a more rigorous, accurate, and comprehensive quality inspection of wire rods. Such an approach not only enhances the quality of the finished product but also can lead to significant economic savings by reducing wastage and rework. It's an embodiment of how traditional industries can be innovatively transformed with the infusion of modern technology.

Acknowledgments. Funding: This work was supported by the Korea Evaluation Institute of Industrial Technology (KEIT) and the Ministry of Trade, Industry, & Energy (MOTIE) of the Republic of Korea [Grant number RS-2022–00155473]: Development of energy efficiency improvement and quality improvement technology by applying big data in the steel rolling process.

References

1. Harste, K., Wustner, E.: Innovative solutions for the production of wire rod and bar. Stahl Eisen **124**(1), 43–48 (2004)
2. Koymatcik, H., Ahlatci, H., Sun, Y., Turen, Y.: Effect of carbon content and drawing strain on the fatigue behavior of tire cord filaments. Eng. Sci. Technol. Int. J. Jestech **21**(3), 289–296 (2018). https://doi.org/10.1016/j.jestch.2018.04.010
3. Pavlina, E.J., Van Tyne, C.J.: Correlation of yield strength and tensile strength with hardness for steels. J. Mater. Eng. Perform. **17**(6), 888–893 (2008). https://doi.org/10.1007/s11665-008-9225-5
4. Wei, Y., et al.: Evading the strength- ductility trade-off dilemma in steel through gradient hierarchical nanotwins. Nat. Commun. **5** (2014). https://doi.org/10.1038/ncomms4580
5. Choi, Y.S., Kim, S.J., Park, I.M., Kwon, K.W., Yoo, I.S.: Boron distribution in a low-alloy steel. Met. Mater. Korea **3**(2), 118–124 (1997). https://doi.org/10.1007/BF03026135
6. Jahazi, M., Egbali, B.: The influence of hot rolling parameters on the microstructure and mechanical properties of an ultra-high strength steel. J. Mater. Process. Technol. **103**(2), 276–279 (2000). https://doi.org/10.1016/S0924-0136(00)00474-X
7. Hwang, J.K.: Effect of ring configuration on the deviation in cooling rate and mechanical properties of a wire rod during the stelmor cooling process. J. Mater. Eng. Perform. **29**(3), 1732–1740 (2020). https://doi.org/10.1007/s11665-020-04694-0
8. Hwang, J.K.: Effects of nozzle shape and arrangement on the cooling performance of steel wire rod in the Stelmor cooling process. Appl. Therm. Eng. **164** (2020). https://doi.org/10.1016/j.applthermaleng.2019.114461
9. Hwang, R., Jo, H., Kim, K.S., Hwang, H.J.: Hybrid model of mathematical and neural network formulations for rolling force and temperature prediction in hot rolling processes. IEEE Access **8**, 153123–153133 (2020). https://doi.org/10.1109/ACCESS.2020.3016725
10. Ronneberger, O., Fischer, P., Brox, T.: U-net: convolutional networks for biomedical image segmentation. In: Navab, N., Hornegger, J., Wells, W.M., Frangi, A.F. (eds.) MICCAI 2015. LNCS, vol. 9351, pp. 234–241. Springer, Cham (2015). https://doi.org/10.1007/978-3-319-24574-4_28
11. Daigo, I., Murakami, K., Tajima, K., Kawakami, R.: Thickness classifier on steel in heavy melting scrap by deep-learning-based image analysis. ISIJ Int. **63**(1), 197–203 (2023). https://doi.org/10.2355/isijinternational.ISIJINT-2022-331
12. Laub, M., et al.: Determination of grain size distribution of prior austenite grains through a combination of a modified contrasting method and machine learning. Praktische Metall.-Pract. Metall. **60**(1), 4–36 (2022). https://doi.org/10.1515/pm-2022-1025

13. Thomas, A., Durmaz, A.R., Straub, T., Eberl, C.: Automated quantitative analyses of fatigue-induced surface damage by deep learning. Materials **13**(15) (2020). https://doi.org/10.3390/ma13153298
14. Xie, L., Li, W., Fan, L., Zhou, M.: Automatic identification of the multiphase microstructures of steels based on ASPP-FCN. Steel Res. Int. (2023). https://doi.org/10.1002/srin.202200204

Industrial Object Detection: Leveraging Synthetic Data for Training Deep Learning Models

Sarah Ouarab[1,3(✉)], Rémi Boutteau[1], Katerine Romeo[1], Christele Lecomte[1], Aristid Laignel[2], Nicolas Ragot[2], and Fabrice Duval[2]

[1] Univ Rouen Normandie, INSA Rouen Normandie, Université Le Havre Normandie, Normandie Univ, LITIS UR 4108, 76000 Rouen, France
sarah.ouarab@hesam.eu
[2] LINEACT CESI, Rouen, France
[3] LINEACT CESI, Lyon, France

Abstract. The increasing use of synthetic training data has emerged as a promising solution in various domains due to its ability to provide accurately labelled datasets at a lower cost compared to manually annotated real-world data. In this study, we investigate the use of synthetic data for training deep learning models in the field of industrial object recognition. Our goal is to evaluate the performance of different models trained with varying ratios of real and synthetic data, with the aim of identifying the optimal ratio that yields superior results. In addition, we investigate the impact of introducing randomisation into the synthetic data on the overall performance of the trained models. The results of our research contribute to the understanding of the role of synthetic data in industrial object detection.

Keywords: Synthetic data · Deep Learning · Detection models · Industrial Objects

1 Introduction

Meanwhile, the automotive industry is in transition to Industry 4.0 with technologies such as the internet of things and big data, Industry 5.0 is already here where mobile robots play an important role and work closely with humans in shared industrial environments. As their responsibilities grow, it becomes imperative for these robots to possess a certain level of autonomy.

One solution to the challenge of creating autonomous robots is the use of state-of-the-art detection models to enable the robot to recognize and localize industrial objects encountered in its path. These models require enormous amounts of data for training, which are often scarce and difficult to obtain due to security and confidentiality reasons, also in some environments data collection can be dangerous. This has led researchers to consider the generation of synthetic data that represents the real-world industrial environment.

S.-H. Sheu (Ed.): ICIEA-EU 2024, LNBIP 507, pp. 200–212, 2024.
https://doi.org/10.1007/978-3-031-58113-7_17

The generation of synthetic data has received a lot of attention from researchers in recent years, and work has been carried out in various domains to fill the lack of real data and enable the use of Deep Learning (DL) algorithms. Some examples are synthetic data generation for deep learning in pedestrian counting [1], synthetic training for real-world object detection [2], and synthetic object recognition datasets for industry [3].

Our contribution focuses on the creation and use of synthetic datasets for the development of high-performance recognition models in the context of industrial objects. These datasets contain randomisation in various aspects, including the colour, rotation and position of the objects, as well as randomisation in the background. By combining the synthetic data generated with the limited real-world data available, we are trying to determine the optimal ratio and parameters for randomisation that will give the most favourable results.

The rest of this paper is organised as follows. In Sect. 2, we present some of the previous work on the use of synthetic data in deep learning model training. Section 3 introduces the synthetic data generation process and provides a detailed explanation of our approach. In Sect. 4, we evaluate the different generated datasets using the YOLOv8 industrial object detection model (Fig. 1).

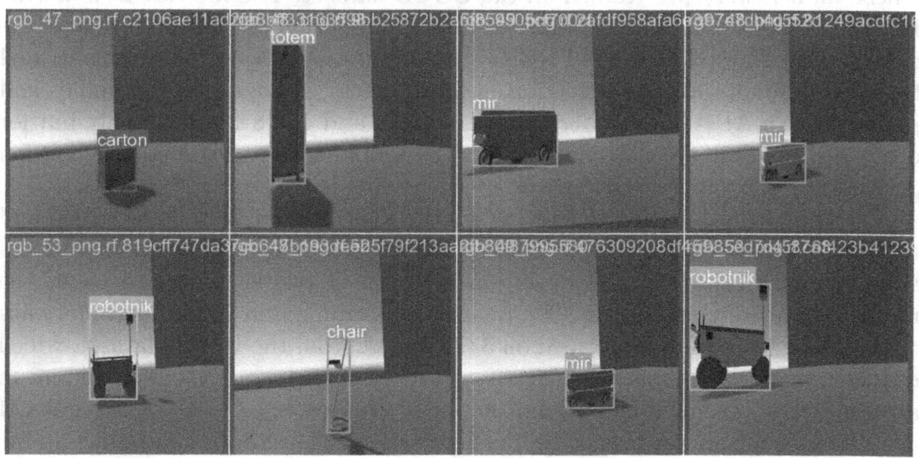

Fig. 1. Industrial Object Detection Enhanced with Synthetic Data

2 Related Work

With the rapid progress in the field of deep learning, many algorithms and models have emerged, especially in the field of computer vision and object detection, where YOLO [4] (You Only Look Once) and SSD [5] (Single Shot MultiBox Detector) are widely used algorithms.

SSD (Single Shot MultiBox Detector) is a widely recognised algorithm in the field of computer vision and object detection. It is known for its impressive speed and accuracy in detecting objects in images. On the other hand, YOLO, which has received considerable attention and recognition for its unique approach to object detection, has released its latest iteration, YOLOv8, and it has gained prominence for its ability to integrate object detection components into a single neural network, resulting in improved performance and flexibility.

Detection models have undergone significant development and have reached high levels of performance. To achieve such performance, it is essential that they are trained on significant amounts of data.

In recent years, with the increasing use of artificial intelligence in various domains, and because in certain cases obtaining a large amount of training data can be a difficult, if not impossible, task. There has been an increasing use of synthetic large-scale labelled datasets to train and optimise deep learning models. In this section, we present some of the previous work on training deep learning models using synthetic data.

In [3] they propose a synthetically generated dataset containing 200,000 images of 8 industrial objects (Cabinet, Stillage, KLT, Box, Jack, Pallet, Pallet, Fire Extinguisher, Smart Transport Robot STR, Dolly) to train a deep learning model using transfer learning. In this work, several models were used, namely FRCNN Resnet 50, FRCNN Resnet-101, SSD Inception and SSD Mobilenet. These models were pre-trained on the COCO dataset. The metrics used are the Intersection over Union (IoU) and the Average Precision (AP). The results of the experiments showed that the FRCNN Resnet-50 model achieved promising and favourable results for the detection of four objects: stillages, STRs, dollies and pallets with a AP@0.5 of 69%, 89.99%, 48.60% and 46.98% respectively but for other objects that are not characterised by a unique texture, shape, dimension or location, the accuracy decreases.

In [1] an algorithm for generating synthetic pedestrian data is proposed and used to generate one million synthetic images. These images are used to train a designed DCNN (Deep Convolutional Neural Network) model to count the number of pedestrians, using 800k images for training and 200k images for validation. The tests were performed on 3375 images from the UCSD dataset, which contains manually labelled images of crowds in pedestrian walkways; the mean absolute error obtained is 1.38 and the mean squared error is 3.6. The results of the experiments indicate that the inclusion of synthetic data can be a suitable alternative to compensate for the absence of real data.

3 Synthetic Data Generation

Synthetic data refers to artificially generated data using computer algorithms or simulations. The use of synthetic data to train deep learning models emerged and quickly gained popularity as a technique to overcome the lack of data and allow model development regardless of the availability of real-world data.

Synthetic data offers many advantages, including the ability to have personalised data, reduced cost and time associated with data collection and annotation. It is easy and quick to obtain. In our situation, we needed images of mobile industrial objects to facilitate their identification by autonomous robots using a DL (Deep Learning) model. In order to ensure accurate and efficient results, this model requires thorough training on a significant amount of data, which we generated using UNITY.

Unity is a real-time development platform that enables the creation of 3D applications, architecture, films and more [6,7]. It comes with a number of packages that refer to a collection of resources that can be easily imported into a Unity project.

Unity was originally known for developing modern games, but with the introduction of a new package called the 'Perception Package', which offers the ability to create large synthetic annotated datasets. Unity has become very popular for data generation.

The Unity Perception Package serves as a comprehensive toolset designed to create large datasets for machine learning tasks related to perception. Its primary focus is to accurately capture ground truth data for camera-based scenarios [8].

In our case, we used Unity's Perception package to create a synthetic database of our industrial objects. Here are the steps we took:

1. **Selection of Industrial Objects:** To begin with, we carefully selected six specific industrial objects for detection, chosen on the basis of their likelihood to be encountered along the robot's path. The objects are visually represented in Fig. 2.

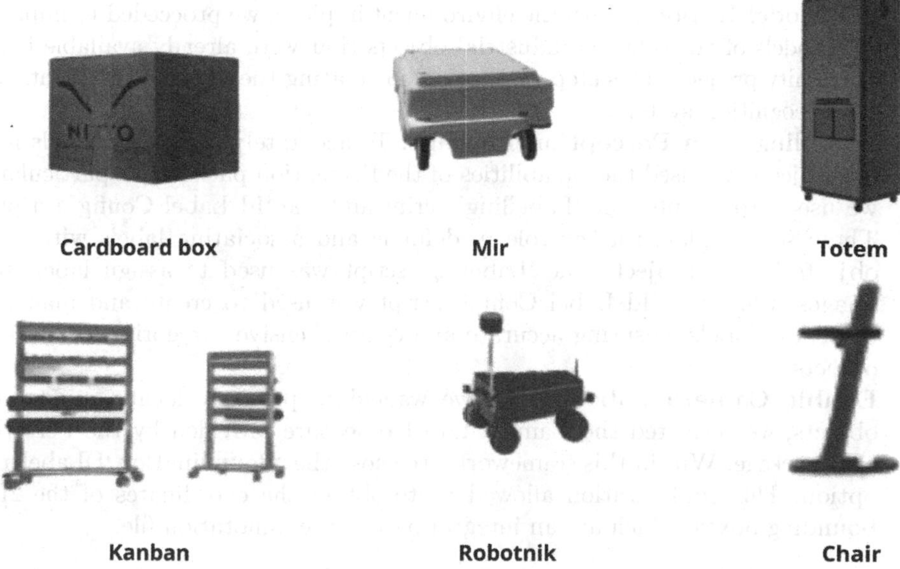

Fig. 2. Industrial Objects to detect

2. **Environment Selection:** We retrieved the 3D model of the laboratory's digital twin, which constitutes our working environment housing industrial objects as well as the robot. The 3D environment is depicted in Fig. 3.

Fig. 3. 3D Digital Twin of the working space

3. **3D Model Import:** With the environment in place, we proceeded to import 3D models of the selected industrial objects that were already available into our Unity project. This step was crucial in creating the visual components of our recognition system.
4. **Labelling with Perception Package:** To accurately define the labels for our objects, we used the capabilities of the Perception package. In particular, we used two scripts: the 'Labelling' script and the 'Id Label Config' script. These scripts played a key role in defining and associating labels with the objects in our project. The 'Labeling' script was used to assign labels to images, while the 'Id Label Config' script was used to create and manage the list of labels, ensuring accurate and comprehensive categorisation of the objects.
5. **Enable Camera Labelers:** As we wanted to precisely locate and track objects, we activated the Camera Labelers feature provided by the Perception package. Within this framework, we chose the "BoundingBox2DLabeler" option. This configuration allowed us to obtain the coordinates of the 2D bounding boxes, which are an integral part of the annotation file.

At the end of these steps, we have a large database containing images of our objects in the environment of our choice.

The Perception package includes a randomiser script that played a crucial role in increasing the diversity of our dataset. This tool allowed us to introduce a wide range of variations, including rotations, colours, backgrounds and more. One of the remarkable features of this script is its flexibility in selecting the specific object or attribute we wanted to randomise.

For example, when it came to manipulating the rotation of objects, we were able to precisely define the axis of rotation, allowing us to create a variety of perspectives for each object in our dataset, ensuring that the variations met the requirements of our project.

Alternatively, when working with background randomisation, we have the flexibility to select the background and choose from different options. We can change the background colour or, if required, import a set of images into Unity to integrate as background elements to enrich the visual experience.

Figure 4 shows some examples of randomisation applied to the Robotnik, (a): Uniform background colour randomiser, (b): Rotation randomiser applied to the object, (c): Camera angle randomiser, (d): Colour randomiser applied to the Robotnik object.

4 Experimentation and Evaluation

In this section on experimentation and evaluation we will present two parts. In the first part, we will perform tests to define the ideal percentage ratio of real and synthetic data that gives the best performance. In the second part, we will keep the percentage that gave the best results and introduce randomisation into the synthetic data, including object colour, rotation and background colour. This will allow us to study the effect of randomisation on the results and further improve performance.

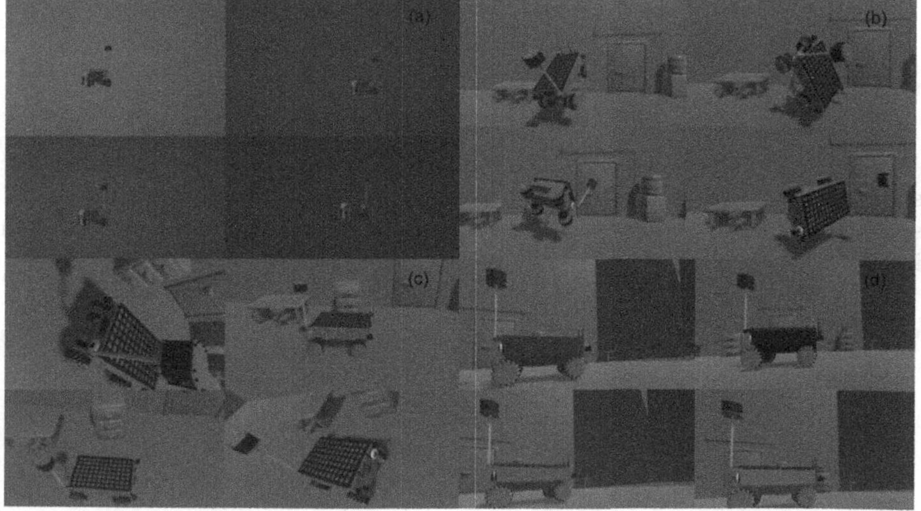

Fig. 4. Perception Package Randomisation On the Robotnik

In the experiments, we used transfer learning by using the pre-trained YOLOv8 model 'YOLOv8n', which was trained on the COCO dataset [9]. For each experiment, we trained our customised detection model for 100 epochs.

The metrics used to evaluate the performance of the models are Mean Average Precision (mAP), which measures the average precision across multiple queries or instances and provides an overall assessment of the accuracy and ranking quality of the system, and F1-score, which considers both precision and recall to provide a balanced assessment and is calculated as the harmonic mean of precision and recall. It is often used when the data is unbalanced and a balance between precision and recall is desired.

4.1 Part 1: Ratio (%) Real/Synthetic

The tests for this part were carried out on six databases of the same size, i.e. 890 images. Table 1 shows the composition of the datasets, with the first database containing only synthetic images. Real images were then gradually added to the databases at a rate of 10% until 50% was reached, in order to observe the effect of adding synthetic data and whether it could potentially solve the problem of data scarcity for training purposes.

Table 1. Datasets' Composition, Ratio(%) Real/Synthetic

Ratio (%) Real/Synthetic	Number of Real-world images	Number of Synthetic images
0R/100S	0	890
10R/90S	89	801
20R/80S	178	712
30R/70S	267	623
40R/60S	356	534
50R/50S	445	445

The performance results of the models trained on these six databases are shown in Fig. 5 for the mAP (mean average precision) and in Fig. 6 for the F1 score. It can be seen that performance improves as the percentage of real data increases.

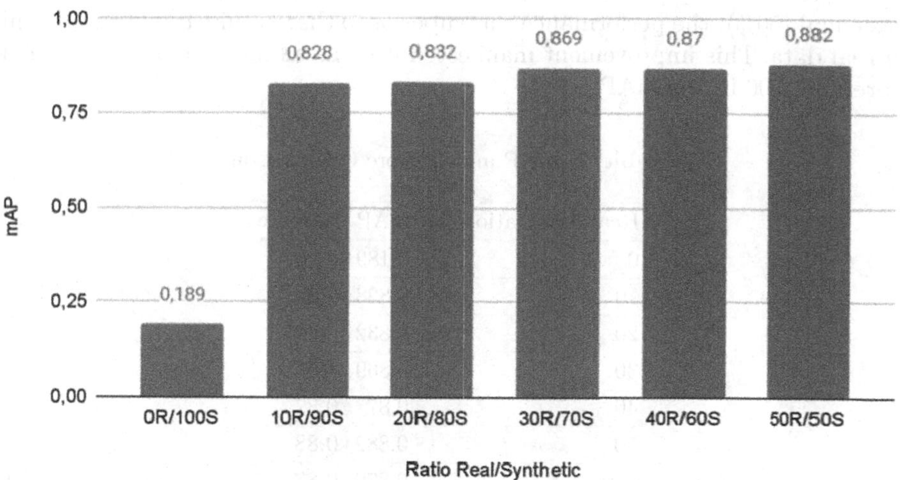

Fig. 5. mAP results.

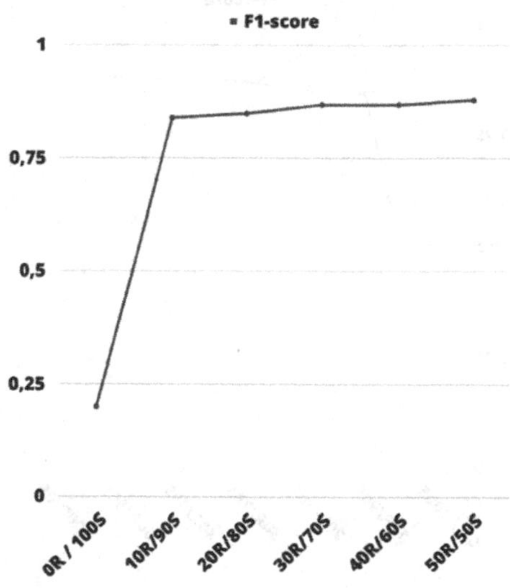

Fig. 6. f1-score results.

For this experiment, our primary objective is to determine the percentage of real and synthetic data that can produce better results than training only on the 445 real images available.

To investigate this question, we compared the results previously obtained by training our models on the six databases with the results obtained by training the model only on the 445 real images.

After analysing the results in Table 2, which are shown in Figs. 7 and 8, we observed that when the proportion of real and synthetic data in the database was equal (50%), the performance was superior to that obtained by training only on real data. This improvement manifested itself as an increase of 0.01 in the f1 score and 0.06 in the mAP.

Table 2. mAP and f1 score Comparison

Real data ratio (%)	mAP	f1-score
0	0.189	0.20
10	0.828	0.84
20	0.832	0.85
30	0.869	0.87
40	0.87	0.87
50	0.882	0.88
100	0.876	0.87

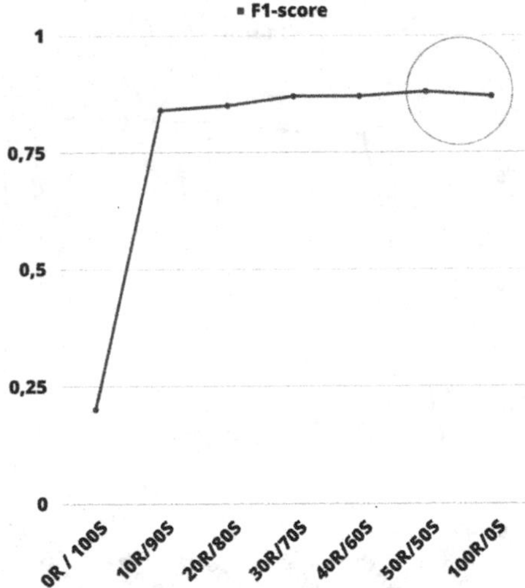

Fig. 7. Comparison between the results of F1 score in Part1's experiment

From these results, we observed that the addition of as little as 10% synthetic data noticeably improved the results. This improvement can be attributed to the fact that the introduction of synthetic data provides new information to the model during training. However, we also found that a substantial increase in the amount of synthetic data did not significantly improve the model's performance. This lack of substantial improvement is due to the repetitive nature of the information in the synthetic data, as we primarily used a single environment. To potentially improve the results further, it may be worth considering incorporating data from different environments or sources.

Furthermore, we also found that we achieved satisfactory mAP and f1 values with only 10% of the real data. This demonstrates that synthetic data significantly fills the gap created by the lack of real data in our scenario, and prompted us to conduct tests to determine the minimum percentage required to achieve a favourable mAP and f1 score, and thus effective detection.

The results of the experiments we conducted to determine this minimum percentage of real data required are shown in the following figures. The mAP and f1 score results are shown in Fig. 9.

In the previous two figures, it is clear that with only 3% real data combined with 97% synthetic data, we obtain satisfactory results, specifically an mAP of 0.733 and an f1-score of 0.73.

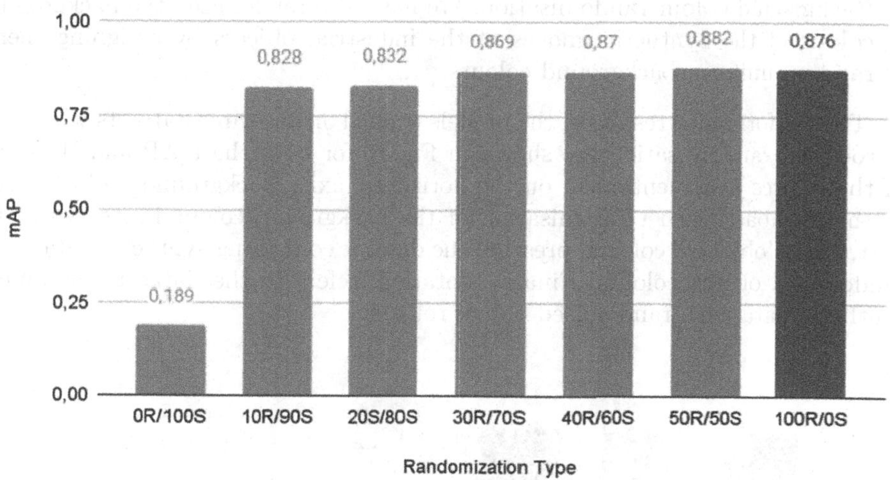

Fig. 8. Comparison between the results of the mAP in Part1's experiment

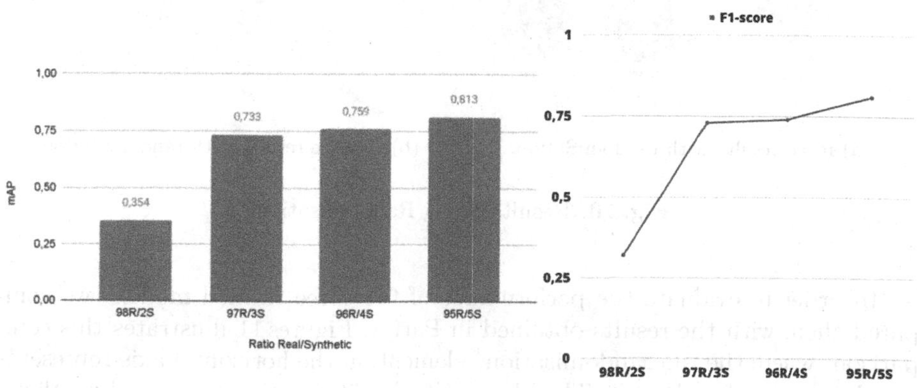

Fig. 9. mAP and f1-score results

4.2 Part 2: Randomisation

In this part, the tests were carried out using datasets with a constant size of 890 and with the same percentage of real and synthetic data. The reason for this is that in the initial testing phase, when we determined the best ratio, the 50% ratio gave the best performance, even outperforming the use of real data only.

The randomiSations that we have introduced to the synthetic data portion are:

- Object colour randomisation: We have applied randomisation to the colours of the industrial objects.
- Object Rotation Randomisation: We have randomised the rotation of the industrial objects by applying random rotations along different axes.
- Background colour randomisation: We have also randomised the background colour of the synthetic images of the industrial objects by assigning them random uniform background colours.

The performance results of the models trained on our three datasets where we introduced randomisation are shown in Fig. 10 for both the mAP and f1-score. In the figures just mentioned, on the horizontal axis, 'background_color' refers to the database with randomisation of the background colour in its synthetic part, while 'objects_color' represents the dataset containing synthetic data with randomised object colours. Finally, 'rotation' refers to the dataset containing synthetic data with randomised object rotations.

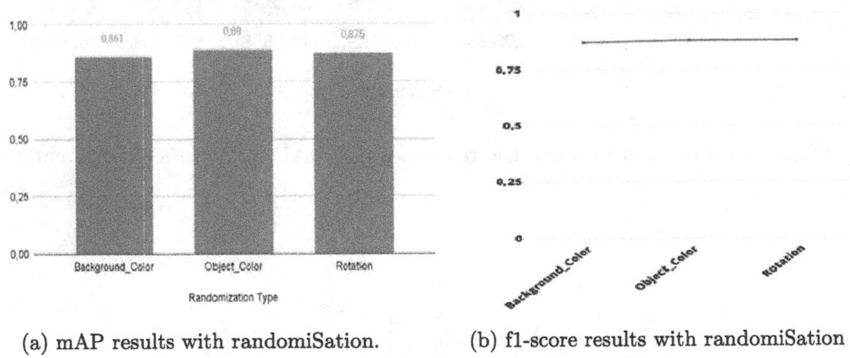

(a) mAP results with randomiSation. (b) f1-score results with randomiSation

Fig. 10. Results Using Randomisation

In order to evaluate the performance of the three trained models, we compared them with the results obtained in Part 1. Figures 11 illustrates this comparison, where the "no randomisation" element on the horizontal axis represents the dataset used in Part 1. This dataset had a 50% ratio of real and synthetic data, with no randomisation applied.

The results show that the integration of different colours of industrial objects into the database resulted in a 0.008 improvement in mAP compared to the database with a 50% ratio of real and synthetic data. There was also an improvement of 0.014 compared to training on real data only. However, no improvement in F1 score was observed.

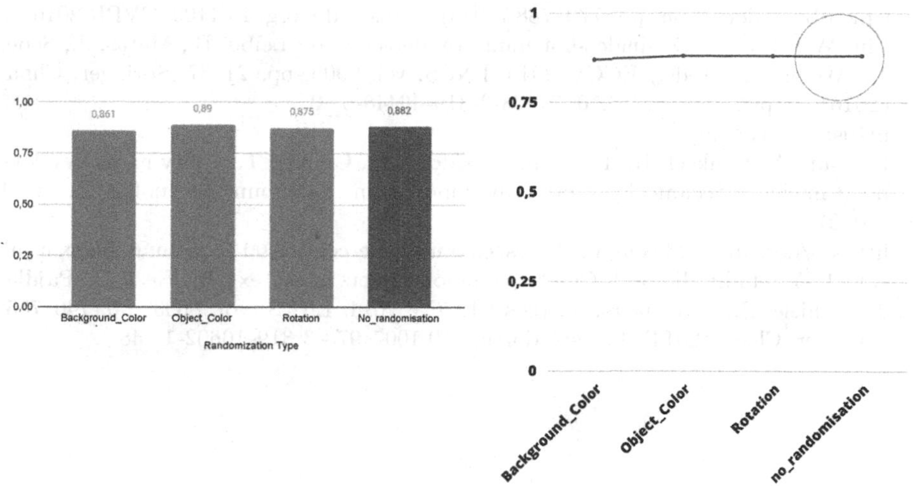

Fig. 11. Comparison between the results of mAP and f1-score in Part2's experiment

5 Conclusion

Based on the various experiments conducted, it is evident that the use of synthetic data in training deep learning models for industrial object detection brings significant benefits. Our results have shown that by achieving a balance of 50% synthetic data and 50% real data, we have been able to outperform the results obtained with real data alone, and that the inclusion of objects with different colours also has a positive impact on the results.

In conclusion, using synthetic data to train deep learning models for industrial object recognition is a beneficial approach that improves results. Combining synthetic and real data in a well-balanced manner leads to improved performance. To further optimise results, ongoing efforts to improve the quality of synthetic data should be pursued, offering the potential for even better results in the future.

References

1. Ekbatani, H., Pujol, O., Seguí, S.: Synthetic data generation for deep learning in counting pedestrians, pp. 318–323 (2017). https://doi.org/10.5220/0006119203180323
2. Gastelum, Z., Shead, T., Higgins, M.: Synthetic training images for real-world object detection (2020)

3. Akar, C.A., Tekli, J., Jess, D., Khoury, M., Kamradt, M., Guthe, M.: Synthetic object recognition dataset for industries. In: 2022 35th SIBGRAPI Conference on Graphics, Patterns and Images (SIBGRAPI), Natal, Brazil, pp. 150–155 (2022). https://doi.org/10.1109/SIBGRAPI55357.2022.9991784
4. Redmon, J., Divvala, S., Girshick, R., Farhadi, A.: You only look once: unified, real-time object detection, pp. 779–788 (2016). https://doi.org/10.1109/CVPR.2016.91
5. Liu, W., et al.: SSD: single shot multibox detector. In: Leibe, B., Matas, J., Sebe, N., Welling, M. (eds.) ECCV 2016. LNCS, vol. 9905, pp. 21–37. Springer, Cham (2016). https://doi.org/10.1007/978-3-319-46448-0_2
6. https://unity.com/fr
7. Hussain, A., Shakeel, H., Hussain, F., Uddin, N., Ghouri, T.: Unity game development engine: a technical survey. Univ. Sindh J. Inf. Commun. Technol. **4**(2), 73–81 (2020)
8. https://docs.unity3d.com/Packages/com.unity.perception@1.0/manual/index.html
9. Lin, T.-Y., et al.: Microsoft COCO: common objects in context. In: Fleet, D., Pajdla, T., Schiele, B., Tuytelaars, T. (eds.) ECCV 2014. LNCS, vol. 8693, pp. 740–755. Springer, Cham (2014). https://doi.org/10.1007/978-3-319-10602-1_48

A Multi-step Approach for Identifying Unknown Defect Patterns on Wafer Bin Map

Jin-Su Shin[1,2] and Dong-Hee Lee[3](✉)

[1] Department of Semiconductor and Display Engineering, Sungkyunkwan University, 2066, Seobu-ro, Suwon-si, Gyunggi-do 16419, Republic of Korea
jinsushin@g.skku.edu

[2] Memory Division, Samsung Electronics Co, Ltd., 1-1 Samsungjeonja-ro, Hwaseong-si, Gyeonggi-do 18448, Republic of Korea

[3] Department of Industrial Engineering, Sungkyunkwan University, 2066, Seobu-ro, Suwon-si, Gyunggi-do 16419, Republic of Korea
dhee@skku.edu

Abstract. In this study, we propose a framework for detecting, classifying, and visualizing unknown patterns in semiconductor wafer defect analysis to improve automation in the field. Rapid advancements in semiconductor processes and equipment have led to the emergence of new defect types, most of which are analyzed and identified based on engineers' experience and judgment. Current approaches struggle with limited labeling, emerging defects, and class imbalance, and although pattern recognition and deep learning techniques have been applied in research, they do not provide a complete solution. We present a method that can quickly detect various emerging defect patterns and ensure high classification accuracy for known defect types. To achieve this, we utilize One Class SVM and Transfer Learning-based ResNet50 backbone, which can be easily implemented on-site. The proposed method uses the one-class SVM method and the validation threshold of each classifier to perform multi-stage unknown defect pattern detection. This approach overcomes the limitations of traditional defect analysis, supporting the identification of new defect types and enhancing engineers' work efficiency. Furthermore, we employ T-SNE and DBSCAN techniques for dimensionality reduction and visualization, providing high accuracy and dimensionality reduction in identifying new defect patterns. These techniques aid engineers in timely labeling and decision-making, ensuring a more efficient response to emerging defects in the semiconductor industry. Consequently, this study offers a comprehensive framework that addresses the challenges of limited labeling, emerging defects, ultimately improving the performance of semiconductor wafer defect analysis. The effectiveness of the proposed model is evaluated through various experiments.

Keywords: Wafer Defect Map Classification · Convolution Neural Network · Density Based Clustering · Unknown Defect Pattern

S.-H. Sheu (Ed.): ICIEA-EU 2024, LNBIP 507, pp. 213–226, 2024.
https://doi.org/10.1007/978-3-031-58113-7_18

1 Introduction

The semiconductor industry is a driving force behind modern technology, with semiconductor chips revolutionizing numerous fields like telecommunications, healthcare, and energy. The industry's progress hinges on shrinking and integrating integrated circuits, boosting processing power. However, this miniaturization faces challenges due to the constant wafer size, leading to complex fabrication processes. Against this backdrop, early detection of wafer defects is crucial to increase the yield of wafers and produce reliable chips.

The Wafer Bin Map (WBM) records and maps test results, aiding in defect identification during manufacturing. Furthermore, by analyzing specific defect types (as in Table 1), it becomes a valuable tool to optimize production processes and improve product quality. Despite this importance, WBM monitoring, defect detection and identification are still performed using individual experience and intuition, relying on engineers' domain knowledge and perspective [1].

To address these challenges and efficiently classify WBM defects, recent advancements in pattern recognition and deep learning image processing technologies have been employed in the wafer map defect classification domain [2]. These technologies aim to automate WBM defect identification and achieve higher classification accuracy. Numerous studies are being conducted to explore and improve these innovative techniques for wafer defect classification.

Table 1. Various WBM defects patterns and causes of defects.

Defect Name	WBM Image	Defect Type Description	Source of defects
Center		Defect patterns concentrated in the center of the wafer	Defects caused by abnormal RF (radio frequency) operation [3] or abnormal fluid flow
Donut		Defect pattern in the form of a circle with a hole in the center	Defect type due to deposition of residues that could not be erased during photoresist cleaning
Edge-Loc		A type of defect pattern clustered on the edge of the wafer	Defects due to abnormal temperature annealing [4] or defects due to valve impurities in the load lock
Edge-Ring		Defective patterns along the rounded edges of the wafer	Defect type due to abnormal temperature control during RTP (Rapid Thermal Process) [5]

(continued)

Table 1. *(continued)*

Defect Name	WBM Image	Defect Type Description	Source of defects
Loc		Defective patterns that occur due to crowding in the wafer	Defective pattern due to vacuum pressure difference due to slit valve leak or poor pump operation, Vibration due to poor fastening of internal parts [3]
Near-Full		Defective patterns across the entire wafer	PR (photoresist) burst due to electronic overcharge on the wafer surface, defect due to abnormal plasma ion beam implant process
Scratch		A type of pattern in which bad patterns are continuously connected within a wafer	During the wafer handling sequence Wafer surface interference by In-vac, ATM Robot, etc., or surface damage by humans, Bad CMP Polishing [6]
Random		Failures occur irregularly with no specific pattern	Irregular defect pattern due to abnormal vacuum or gas flow

However, most WBM image classification studies mainly focus on applying state-of-the-art deep learning models to improve the performance of supervised learning using correct answer labels from existing data [1]. To compensate for the disadvantages of this approach, an unsupervised learning technique that assumes that all data do not have correct labels is also being actively studied [7]. In addition, various semi-supervised learning techniques that define and learn new labels based on limited label information have recently been applied and evaluated in research [8].

The aim of this study is to quickly detect various unknown defect patterns and achieve high classification accuracy for known defect types by overcoming the limitations of traditional learning methods such as supervised learning, unsupervised learning, and semi-supervised learning. To this end, this study introduces a multi-step methodology for detecting unknown patterns using a one-class SVM and classifier thresholds for each class in a pre-trained ResNet50 model. The proposed method overcomes the limitations of conventional defect analysis, avoids false recognition of new defect types by detecting and visualizing newly created defect types, and enables engineers to quickly identify and label new defects. At the same time, it provides high-precision classification performance for known defect patterns. This paper reviews various studies on Wafer Bin Map (WBM) classification in Sect. 2, describes the multistep methodology for unknown pattern detection proposed in Sect. 3, and the modules used. In Sect. 4, the performance of the proposed network is shared in detail through various evaluations, and finally, in Sect. 5, conclusions and future research directions are presented.

2 Related Work

As a large amount of data is secured and the performance of various processing units improves, AI technology naturally develops and is in the spotlight. Naturally, research on the application of AI and deep learning technologies is being actively conducted in various manufacturing and production process fields, and it is also being rapidly applied in the semiconductor industry, which requires strict reliability for optimal production efficiency and quality. In particular, the latest image processing deep learning technology based on CNN is in the limelight in the field of WBM image classification, which can quickly determine the suspect process by identifying the cause of wafer defects. In this section, various related works applied to WBM research are analyzed and research trends are described.

2.1 Supervised Learning

For most wafer bin map (WBM) studies, supervised learning is applied to identify defect locations by learning both good and bad patterns within the WBM. Table 2 highlights research efforts in this area, aiming to enhance the accuracy and speed of WBM classification using modern models that have emerged with technological advancements. In Qiao Xu et al., 2022 [9], to improve the classification performance of ResNet18 using the residual module, an Improved CBAM module was applied to improve the data imbalance problem and classification performance. Likewise, Cha J, Jeong., In 2022 [10], classification performance was evaluated by applying the CBAM module based on U-Net. Chen S, Zhang Y., 2022 [2] also proposed a study to classify WBM using DCNN (Deep Convolutional Neural Networks) as a classification technique for WBM. In addition, Subhrajit et al., 2022 [11], who proposed a WSCN (WaferSegClassNet) network using pixel-level segmentation to maximize classification performance, also conducted supervised learning-based research on WBM classification. Shinde P, Pai P, Adiga S., 2022 [12] authors, since existing WBM classification studies focus only on the classification of defect types, the object detection function is used to predict information on the location where defects occur and utilize the information. They proposed a network using the YOLO (YouOnlyLookOnce) algorithm as a backbone that supports. The predominant focus of these studies revolves around enhancing the classification performance of known defect types within a given Wafer Bin Map (WBM) using state-of-the-art supervised models, often prioritizing swifter classification speeds. Nevertheless, this approach necessitates accurate labeled data for training, which can pose challenges in the semiconductor industry due to issues such as data imbalances among defect types and limitations in labeling all defect types due to random sampling constraints.

2.2 Unsupervised Learning

Unsupervised learning has gained significant attention in addressing the limitations of supervised techniques in WBM misclassification due to the challenge of limited data labeling. Unsupervised learning involves identifying patterns or structures in data without relying on accurate labels for all data points. Techniques like clustering and dimensionality reduction are commonly employed in this approach. Lee J, Moon I, Oh R.,

2021 [7] proposed a model for all types of unlabeled defects using DPGMM (Dirichlet-ProcessGaussianMixtureModel) model to cluster each WBM map and apply similarity priority to other defects, which can be used for new defects.

In unsupervised learning, the learning process and model construction occur without assuming the presence of accurate labels for all data. This can result in slower classification accuracy and speed compared to supervised techniques. Another drawback is that engineers must manually inspect numerous target samples. This verification is essential to validate the clustering or classification of defects and assess the outcomes of newly generated labels.

2.3 Semi-supervised Learning

Semi-supervised learning is a useful approach for scenarios with limited labeled data, gradually generating new labels for unlabeled data by assessing their similarity to labeled examples. In WBM classification research, it's employed to generate and learn new labels for failures resembling existing types, accommodating the possibility of unknown failure types. For instance, in the study by Lee H and Kim H, conducted in 2020 [8], they established 16 latent defect labels by combining four predefined basic defect patterns: circle, scratch, partial ring, and local area. They proposed a model that generates and learns these labels through estimation techniques. However, semi-supervised techniques face a limitation when it comes to detecting entirely new types of failures. This challenge arises because these techniques generate new labels by assessing the similarity of unlabeled data to existing labeled failures.

2.4 Advanced Techniques for WBM Analysis

With the rapid advancements in deep learning, various techniques, including supervised, unsupervised, and semi-supervised learning, have been applied to the field of WBM defect discrimination research. However, each of these techniques has its own limitations. To overcome these limitations, additional studies have been conducted. For example, the "Bin2Vec" technique has been introduced to WBM, which adds color labels to defect images, improving visualization [13]. Additionally, diverse data augmentation techniques have been used to address data imbalance issues [6]. Efforts have also been made to detect out-of-distribution (ODD) defects using VGGNet, aiding in identifying incorrectly classified defects [14]. However, a key drawback of these approaches is the need to simultaneously train for both in-distribution (IND) defects of interest and ODD defects, which poses a challenge. In summary, despite the impressive progress in deep learning technology, addressing the rapid identification of defects and their causes in WBM, especially for previously unknown defect types, remains a challenge. This study aims to create a comprehensive model that addresses the strengths and limitations of each approach, providing a more holistic solution.

Table 2. Related Works for WBM Classification.

Objective	Author	Method	Data Set
Supervised	Qiao Xu et al., 2022 [9]	ResNet 18 (With Improved CBAM)	WM-811K
	Subhrajit et al., 2022 [11]	WSCN (WaferSegClassNet)	WM38 (Mixed Type)
	Chen S, Zhang Y., 2022 [2]	DCNN	WM-811K
	Shinde P, Pai P, Adiga S., 2022 [12]	YOLO V4	WM-811K
	Cha J, Jeong., 2022 [10]	U-Net (With CBAM)	WM-811K & WM38 (Mixed)
Unsupervised	Lee J, Moon I, Oh R., 2021 [7]	Clustering (DPGMM)	WM-811K, Industry Private data
Semi Supervised	Lee H, Kim H., 2020 [8]	SS-CDGMM	Circle, Scratch, Partial Ring, Local Zone, and Mixed Pattern
Wafer Bin Coloring	Kim J, Kim H., 2019 [13]	Bin2Vec (Bin Coloring)	27,071WBM (DRAM, SSD, etc.)
Augmentation	Kim E, Choi S., 2021 [6]	ResNet50	1238 Wafers from industry
Detect Out-Of-Distribution	Kim Y, Cho D, Lee J., [14]	VGGNet	11,789 13 pattern in-distribution 1899 out of-distribution

3 Methodology

In this section, we outline a comprehensive approach to identifying and visualizing unknown defects in test data while improving the identification of known defect types. First, all known defect types are combined into a single dataset, and classification criteria for each class are trained using the One-Class SVM model. Additionally, we use a multi-label classification approach based on the ResNet50 model with transfer learning (see Fig. 1). If the learned thresholds for all classes are not met, the case is classified as an unknown defect, resulting in a multi-stage classification process. Figure 3 demonstrates the entire process of verifying test WBM images with potentially unknown defects using the proposed network. Data classified as unknown defects are further analyzed through dimensionality reduction and density-based DBSCAN clustering.

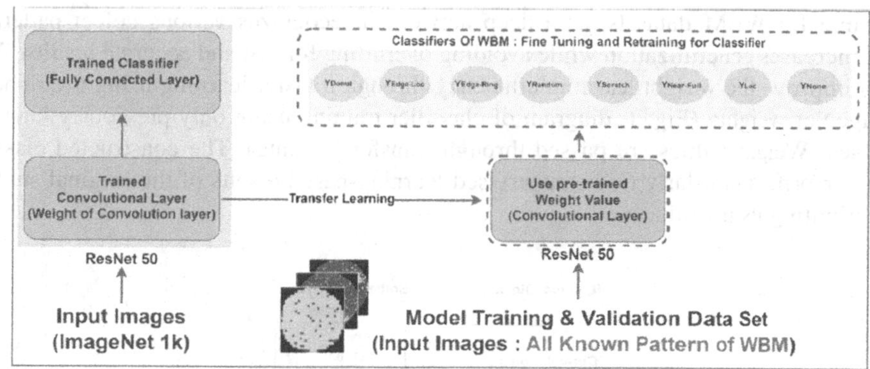

Fig. 1. Supervised classifier configuration diagram for each class of ResNet50 using pre-trained weight values by transfer learning

3.1 One Class SVM

The One-Class SVM is an unsupervised algorithm that detects outliers, creating a boundary between normal and abnormal data in a dataset. It learns normal patterns without being influenced by minority data. When applied to a unified dataset from a WBM study, when applied to an integrated dataset of WBM studies, you can create hyperplane boundaries to classify test data into known and unknown classes. This method outperformed Autoencoder for pattern recognition on WBM data, indicating its effectiveness. Grid Search was used to find optimal hyperparameters, revealing that an NU value of 0.15 and the Radial Basis Function (RBF) kernel function were most effective. The optimized One-Class SVM showed high discriminative power on the WBM dataset, contributing to the successful construction of a virtual hypersurface boundary between normal and outlier data.

3.2 ResNet50 with Transfer Learning

In this study, we used the ResNet50 backbone to create a classifier for each class of previously known patterns to classify known defect patterns in the wafer bin map (WBM). We add a classifier for the Unknown class, a previously undefined defect type, to classify samples that do not meet the training threshold of each classifier as unknown defects. As a result, these features add the ability to determine unknown defects at multiple stages along with One-class SVM. (In the 'Predict for Known Label' branch in Fig. 3, the 'unknown' class is additionally displayed at the top of the classifier) The core deep learning architecture, ResNet50, uses skip connections to overcome vanishing gradients (Fig. 2, left). The bottleneck block structure prevents overfitting by efficiently managing parameters. Despite the depth, ResNet50 maintains high representation learning compared to ResNet18 and ResNet34 (Fig. 2, right). These properties make ResNet50 suitable for WBM datasets, as a deep network it recognizes various defect patterns and increases generalization while avoiding overfitting for fast and accurate results. We also improve the weight learning efficiency through transfer learning using pretrained ImageNet weights (Fig. 1, diagram of classifier training using only previously known classes. Weight values are passed through transfer learning). The constructed classifier performed similarly to the supervised learning-based results of the original study, highlighting its usefulness.

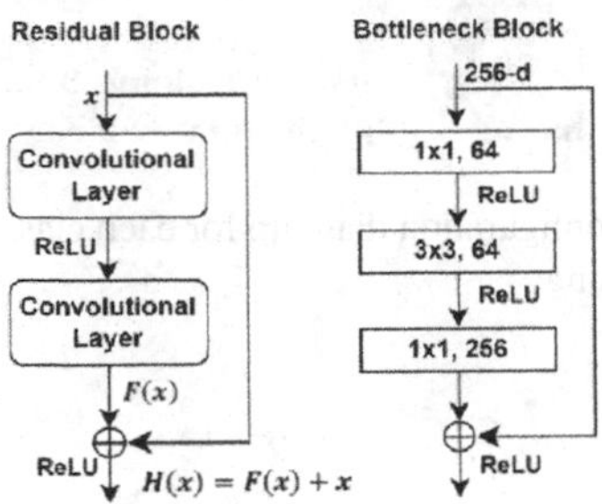

Fig. 2. Left: structure of residual block for ResNet Right: structure of bottleneck block for ResNet50.

3.3 T-SNE & DBSCAN (Clustering)

Detected unknown defect types undergo conditionality reduction and visualization using T-SNE, which preserves data structure while mapping high-dimensional data into a low-dimensional space. The heavy-tailed distribution of T-SNE effectively maintains the distance between data points, allowing for maintenance and visualization of clear cluster boundaries. This approach considers both known and potential unknown defects through a multilevel classification. We also use DBSCAN, a flexible density-based clustering algorithm, to cluster samples identified as unknown defects after conditionality reduction with T-SNE. DBSCAN adapts to data distributions suitable for non-linear cluster identification. This is suitable for the semiconductor industry where the defect number is not known in advance and miss classification, or noise may occur. Figure 3 shows test data validation and shows the clustering results of unknown defect types classified by DBSCAN after conditionality reduction using T-SNE, and finally presented as a wafer map. This approach alerts engineers to new defect occurrences and enables timely decisions through methods such as Pseudo-labeling. Mitigate potential risks in the semiconductor industry by more efficiently responding to unknown failures.

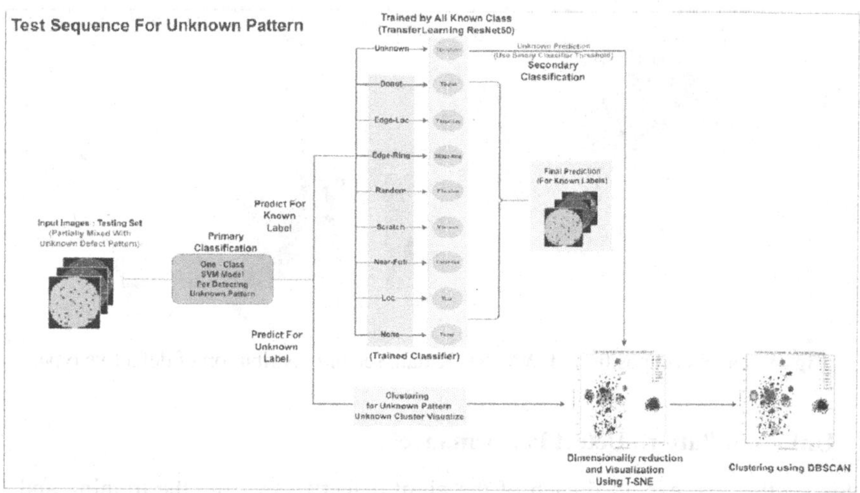

Fig. 3. Multi Step Classification and Visualize Sequence for Potentially Containing Unknown Defect Patterns

4 Experiments

4.1 Data Set: WM-811K

The WM-811K dataset contains a single "normal" pattern and eight categories of "defect" patterns, including Center, Donut, Edge-Ring, Edge-Loc, Loc, Scratch, Random, and Near-full. In total, this consists of 811,457 data points, with approximately 21.3% labeled and the remaining 78.7% unlabeled (see Fig. 4). In this study, we systematically removed each of the eight defect patterns from the dataset to consider the possibility of unknown defect types occurring alongside known defect types. We then split the dataset into 60% training data, 20% validation data, and 20% test data, and utilize only data with known defect pattern classes for pre-training and validation. Additionally, we built an 'Unknown Pattern Mixed Test Dataset' by combining previously excluded defect types with existing test data to assume unknown defects. After performing multi-step detection of unknown defect patterns using test data, we use T-SNE to reduce the predicted samples of unknown classes to lower dimensions to visualize the data. Afterwards, the results of T-SNE are clustered into various clusters using DB-SCAN, and the Wafer Map of the clustered samples is visualized. This procedure allowed us to systematically investigate and visualize the distribution and relationships among the identified unknown defect patterns. On the other hand, samples predicted to exhibit known patterns through a multi-step validation process were classified using a pre-trained classifier trained only on known defect types. This allowed us to evaluate the classifier's ability to accurately classify known patterns within the dataset. In summary, our comprehensive approach evaluates the performance of the model in identifying and classifying both known and unknown defect patterns in the WM-811K dataset.

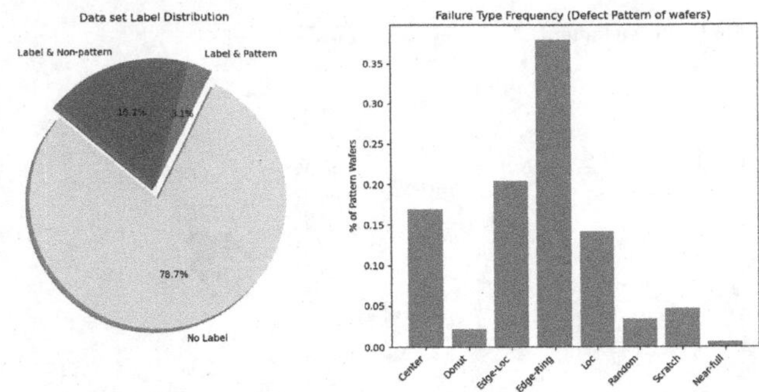

Fig. 4. Label composition of WM-811K data set and distribution of defective type.

4.2 Unknown Pattern Detect Performance

In this section, we excluded each of the eight defect types from the training and validation stages of the dataset as described previously. The excluded defects were then incorporated into a separate test dataset to evaluate the model's performance in identifying and classifying untrained, non-predefined defects. The results of performance evaluation are shown in Table 5, and detailed formulas are shown in Tables 3 and 4. The 'Detect Accuracy' metric is related to the accuracy of detecting unknown defect patterns during a multi-step detection process, while the 'Detect Ratio' metric represents the recall value during a multi-step detection process of an unknown class. We also comprehensively evaluated the overall performance for unknown class detection using 'Weighted F1 Score'. The 'Weighted' values in Table 4 correspond to the number of samples in each class. Since the approach in this study includes multi-step detection utilizing a ResNet50-based classifier for each class in addition to One-Class SVM, we used Weighted F1 Score for evaluation to ensure accurate detection performance for unknown samples. This is especially important because the number of unknown pattern classes compared to known pattern classes may be extremely limited or even non-existent. We also utilize the 'Cluster Proportion' metric to determine whether T-SNE and DBSCAN can cluster unknown defect patterns. The higher the cluster proportion, the more accurate the clustering is and confirms that clusters containing unknown patterns have been successfully formed. In summary, our study demonstrates the effectiveness of the proposed method in identifying and appropriately clustering previously undefined (or untrained) error types when they occur (Fig. 5).

Table 3. Index of confusion matrix.

Confusion Matrix Index		Predicted Class	
		Negative (0)	Positive (1)
Actual Class	Negative (0)	TN (True Negative)	FP (False Positive)
	Positive (1)	FN (False Negative)	TP (True Positive)

Table 4. Formulas for each indicator used to verify the performance of the model.

Index Name	Equation
Detect Accuracy	$\frac{TP+TN}{Total\ Samples}$
Detect Ratio (Recall)	$\frac{TP}{TP+FN}$
Weighted F1 Score	$\frac{2\cdot(Precission_{weighted}\cdot Recall_{weighted})}{Precision_{weighted}+Recall_{weighted}}$
Cluster Proportion	$\frac{Unknown\ Samples\ in\ Cluster}{Total\ Number\ of\ Samples\ in\ cluster}$

Table 5. Performance of Detecting Unknown Patterns in Test Data After Excluding Each Defect Type During Classifier Training.

Unknown Pattern	Detect Ratio	Detect Accuracy	Weighted F1 Score	Cluster Proportion
Center	69	83	85	99
Donut	72	86	91	90
Edge-Loc	42	77	78	83
Edge-Ring	36	70	67	100
Loc	32	79	81	72
Random	99	86	91	100
Scratch	29	82	87	58
Near-Full	100	86	92	100
Average (Macro)	**59.9**	**81.1**	**84**	**87.8**

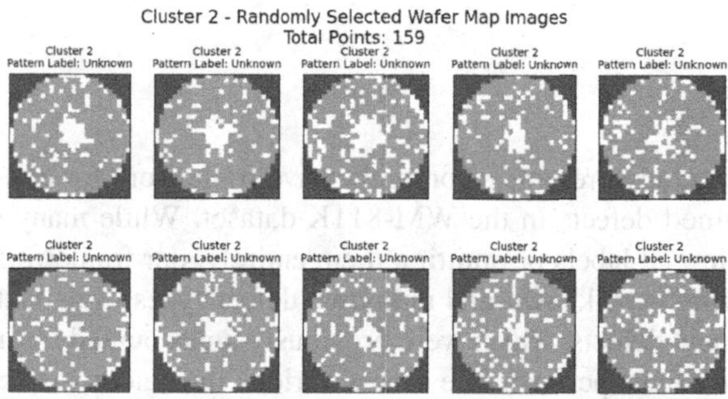

Fig. 5. WBM image of unknown defect cluster formed from test data after assuming 'Center' class as Unknown class.

4.3 Classifier Performance

Table 6 presents the classification results of the test data, including potentially unknown types of defects. Similarly, considering data and class imbalance, we evaluated performance based on overall accuracy and weighted precision, recall, and F1 scores, respectively. As a result, this study can accurately discriminate undefined types, and the discrimination accuracy for predefined types is not significantly different from other recent studies. As a result, we designed an overall framework that uses thresholding and single-class SVM models for classifiers trained only with known failures, assuming new types of failures not covered in related studies, and the model performs well with several metrics.

Table 6. Classifier overall average performance across all classes when tested after assuming each class is an Unknown Class.

Unknown Pattern	Accuracy	Weighted Precision	Weighted Recall	Weighted F1 Score
Center	98.5	93	93	92
Donut	99.6	98	98	98
Edge-Loc	97	88	88	84
Edge-Ring	99	82	76	72
Loc	98	91	91	88
Random	99	99	99	99
Scratch	99	94	94	94
Near-Full	99.7	98	98	98
Average (Macro)	**98.8**	**92.9**	**92.1**	**90.6**

5 Conclusion

In conclusion, our study presents a comprehensive framework for effectively detecting previously undefined defects in the WM-811K dataset. While many studies focus on either incorporating all labels or omitting them entirely, our framework achieves a balance by considering both known and unknown defect types. This method enables the identification of new defects, which were previously unaccounted for in related studies, and demonstrates robust performance across various indicators such as discrimination accuracy and cluster formation. The validation results not only showcase the model's ability to discern previously unknown defects but also emphasize its performance, which is on par with the classifiers designed for known defect types. We have identified potential future directions for enhancing our research. The active learning approach, incorporating Pseudo-Labeling and retraining of the classifiers, promises to further improve our model's accuracy in dealing with unknown patterns. Additionally, by exploring alternative backbone algorithms and incorporating state-of-the-art techniques, we aim to strengthen the threshold accuracy and enhance the classifier's ability to discriminate similar defect types. Overall, our study offers a novel and effective solution for the semiconductor manufacturing industry, providing engineers with the means to rapidly identify emerging defect types. This proactive approach, bolstered by our multi-stage framework, demonstrates the potential to mitigate risks posed by previously unrecognized defects, thus contributing to more efficient and reliable semiconductor production processes.

References

1. Hsu, C.Y., Chen, W.J., Chien, J.C.: Similarity matching of wafer bin maps for manufacturing intelligence to empower Industry 3.5 for semiconductor manufacturing. Comput. Ind. Eng. **142**, 106358 (2020). https://doi.org/10.1016/j.cie.2020.106358
2. Chen, S., Zhang, Y., Hou, X., Shang, Y., Yang, P.: Wafer map failure pattern recognition based on deep convolutional neural network. Expert Syst. Appl. **209**, 118254 (2022). https://doi.org/10.1016/j.eswa.2022.118254
3. Hansen, C.K., Thyregodb, P.: Use of wafer maps in integrated circuit manufacturing. Microelectron. Reliab. **38**(6–8), 1155–1164 (1998)
4. Hansen, M.H., Nair, V.N., Friedman, D.J.: Monitoring wafer map data from integrated circuit fabrication processes for spatially clustered defects. Technometrics **39**(3), 241–253 (1997)
5. Tello, G., Al-Jarrah, O.Y., Yoo, P.D., Al-Hammadi, Y., Muhaidat, S., Lee, U.: Deep-structured machine learning model for the recognition of mixed-defect patterns in semiconductor fabrication processes. IEEE Trans. Semicond. Manuf. **31**(2), 315–322 (2018). https://doi.org/10.1109/TSM.2018.2825482
6. Kim, E.S., Choi, S.H., Lee, D.H., Kim, K.J., Bae, Y.M., Oh, Y.C.: An oversampling method for wafer map defect pattern classification considering small and imbalanced data. Comput. Ind. Eng. **162**, 107767 (2021). https://doi.org/10.1016/j.cie.2021.107767
7. Lee, J.H., Moon, I.C., Oh, R.: Similarity search on wafer bin map through nonparametric and hierarchical clustering. IEEE Trans. Semicond. Manuf. **34**(4), 464–474 (2021). https://doi.org/10.1109/TSM.2021.3102679
8. Lee, H., Kim, H.: Semi-supervised multi-label learning for classification of wafer bin maps with mixed-type defect patterns. IEEE Trans. Semicond. Manuf. **33**(4), 653–662 (2020). https://doi.org/10.1109/TSM.2020.3027431
9. Xu, Q., Yu, N., Essaf, F.: Improved wafer map inspection using attention mechanism and cosine normalization. Machines **10**(2), 146 (2022). https://doi.org/10.3390/machines10020146
10. Cha, J., Jeong, J.: Improved U-Net with residual attention block for mixed-defect wafer maps. Appl. Sci. **12**(4), 2209 (2022). https://doi.org/10.3390/app12042209
11. Nag, S., Makwana, D., Sai Chandra Teja, R., Mittal, S., Mohan, C.K.: WaferSegClassNet - a light-weight network for classification and segmentation of semiconductor wafer defects. Comput. Ind. **142**, 103720 (2022). https://doi.org/10.1016/j.compind.2022.103720
12. Shinde, P.P., Pai, P.P., Adiga, S.P.: Wafer defect localization and classification using deep learning techniques. IEEE Access **10**, 39969–39974 (2022). https://doi.org/10.1109/ACCESS.2022.3166512

13. Kim, J., Kim, H., Park, J., Mo, K., Kang, P.: Bin2Vec: a better wafer bin map coloring scheme for comprehensible visualization and effective bad wafer classification. Appl. Sci. **9**(3), 597 (2019). https://doi.org/10.3390/app9030597
14. Kim, Y., Cho, D., Lee, J.-H.: Wafer map classifier using deep learning for detecting out-of-distribution failure patterns. In: 2020 IEEE International Symposium on the Physical and Failure Analysis of Integrated Circuits (IPFA), Singapore, pp. 1–5 (2020). https://doi.org/10.1109/IPFA49335.2020.9260877

Synthetic Datasets for 6D Pose Estimation of Industrial Objects: Framework, Benchmark and Guidelines

Aristide Laignel[1(✉)], Nicolas Ragot[2], Fabrice Duval[1], and Sarah Ouarab[3]

[1] LINEACT CESI, Rouen, France
aristide.laignel@outlook.fr, fduval@cesi.fr
[2] LINEACT CESI, Caen, France
nragot@cesi.fr
[3] LITIS UR 4108, 76000 Rouen, France
s.ouarab@esi-sba.dz
https://lineact.cesi.fr/

Abstract. This paper falls within the industry 4.0 and tackles the challenging issue of maintaining the Digital Twin of a manufacturing warehouse up-to-date by detecting industrial objects and estimating their pose in 3D, based on the perception capabilities of the robots moving all along the physical environment. Deep learning approaches are interesting alternatives and offer relevant performances in object detection and pose estimation. However, they meet the requirement of large-scale annotated datasets for training the models. In the industrial and manufacturing sectors, these massive datasets do not exist or are too specific to particular use-cases. An alternative aims to use 3D rendering software to build annotated large-scale synthetic datasets. In this paper, we propose a framework and guidelines for creating synthetic datasets based on Unity, which allows the 3D-2D automatic object labeling. Then, we benchmark several different datasets, from planar uniform background to 3D contextualized Digital Twin environment with or without occlusions, for the industrial cardboard box detection and 6D pose estimation based on the YOLO-6D architecture. Two major results arise from this benchmark: the first underlines the importance of training the deep neural network with a contextualized dataset according to the targeted use-cases to achieve relevant performances; the second highlights that integrating cardboard box occlusions in the dataset tends to degrade the performances of the deep-neural network.

Keywords: Synthetic dataset · Industrial object · 6D detection and pose estimation

1 Introduction

The 4th industrial revolution is strongly affected by the massive use of the emerging information and communication technologies such as Robotics, Artificial Intelligence (AI) and eXtended Reality (XR) in pair with the Digital Twin

S.-H. Sheu (Ed.): ICIEA-EU 2024, LNBIP 507, pp. 227–241, 2024.
https://doi.org/10.1007/978-3-031-58113-7_19

(DT), defined as the high-fidelity virtual model of the physical system in virtual space. One current challenge is to keep the DT up-to-date according to the evolutions of the physical system so as the supervisor decision-making would be the most relevant for enhancing productivity, making individualized and lower-cost products, strengthening the operator's safety and efficiency and making better use of available resources [11]. To achieve this physical to virtual updating, one solution consists in taking advantage of the robot perception capabilities by detecting and estimating the 6D pose ($[T_x, T_y, T_z], [R_x, R_y, R_z]$) of the industrial objects disseminated in the warehouse. This framework relies on the use of efficient Deep Learning (DL) approaches, which are currently providing the most relevant results in the literature. Nevertheless, to achieve excellent performances, massive annotated datasets are required for training the models. This is a critical challenge in the industrial and manufacturing sectors, where datasets are rare and limited to specific use-cases. To overcome this issue, an alternative deals with the use of synthetic datasets built from 3D rendering software, which allows generating massive and labeled dataset instantaneously.

This paper falls within this field of creating synthetic labeled datasets for industrial object detection and pose estimation to compensate the data rarity in the manufacturing sector, by proposing:

1. A framework for generating synthetic annotated industrial object datasets based on the Unity 3D rendering software.
2. A comparative study of synthetic datasets for industrial object detection and pose estimation based on YOLO-6D [18].
3. Guidelines for the creation of synthetic labeled datasets to detect and estimate the 6D pose of industrial objects. Which highlights the added-value of a synthetic dataset based on the digital twin of the manufacturing layout.

The paper is organized as follows: Sect. 2 provides an overview of the literature about 6D pose estimation approaches and related datasets. Section 3 proposes a framework for creating synthetic annotated datasets. Section 4 outlines the experimental set-up. Section 5 depicts and analyzes the results of the comparative study for industrial object detection and pose estimation based on YOLO-6D according to different synthetic annotated datasets. Guidelines for creating synthetic datasets according to the specific use-case of detecting and estimating the 6D pose of industrial objects are also provided. Section 6 of this paper provides a conclusion and outlines some perspectives.

2 Related Work

2.1 Methods

3D pose estimation of rigid objects is a well-known problem in computer vision, and several relevant surveys have already been published. In 2016, Marchand et al. [15] propose a literature study about 3D pose estimation for Augmented Reality applications. More recently, Chen et al. [4] draw up an inventory of 6D pose estimation approaches by proposing a classification in 3 main groups:

- *Traditional methods* which are based on: 1) point-pair features such as SIFT, SURF and ORB. As pointed out by Pitteri [16], these approaches provide good performances for textured objects, but fail for textureless ones which is mainly the case of industrial objects; 2) template matching as proposed by Hinterstoisser et al. [7] which is an interested alternative to point-pair features. Nevertheless, this approach requires generating prior massive templates, and is not robust to object occlusions; 3) 3D local features as proposed by Buch et al. [2] which consist in proposing a new pose and voting approach based on the assumption that corresponding oriented points between two models can be used to cast a constrained number of votes for the correct pose aligning the two models.
- *Deep Learning methods* based on Convolutional Neural Networks (CNNs) such as, SSD-6D [13], PoseCNN [20], Deep-6DPose [6], and a real-time deep framework [21]. These approaches are more and more popular in the scientific community since they provide relevant performances in comparison with traditional methods.
- *Mixed approaches* combining traditional and deep learning methods which provide very competitive performances. In 2017, Rad et al. proposed BB8 [17] which uses a CNN to first predict the 2D locations of the projections of the 3D object bounding box and then compute the 6D pose parameters using the Perspective-n-Point (PnP) method. In 2018, Tekin et al. introduced YOLO-6D, a two-stage pipeline mixing 2D keypoints extraction based on a deep learning framework, associated with a 2D-3D equivalent-based PnP mapping. This approach has been upgraded in 2020 by Kang et al. [12] with YOLO-6D+, a single-shot CNN which uses privileged silhouette information which predicts the 2D projections of the 3D bounding box vertices and then estimate the object 6D pose with a PnP algorithm.

2.2 Datasets

Whether it deals with *Deep Learning methods* or *Mixed approaches*, one challenging issue for deploying these methods in the industrial and manufacturing sectors relies on the availability of large-scale annotated real datasets for training the CNN models. Some are available, but they are often specific to particular use-cases. One can refer to: LineMOD Dataset [8] proposed by Hinterstoisser et al.; Occluded-LineMOD Dataset proposed by Brachmann et al. [1] defined as an extended version of LineMOD dataset with multiple occluded objects; YCB Dataset proposed by Calli et al. [3]; T-LESS Dataset from Hodaň et al. [10] which is an RGB-D dataset of texture-less objects; Synthetic T-LESS Dataset proposed by Pitteri et al. which provides photorealistic synthetic images from the T-LESS objects. Recently, Vanherle et al. [19] have provided guidelines for generating appropriate synthetic datasets for training deep learning models to achieve relevant performances towards a generalization to real data. Experiments were conducted with the DIMO dataset [5]. In 2023, Ljungqvist et al. [14] discussed how object detection models are affected when trained on synthetic data versus real data by exposing the inner workings of the network.

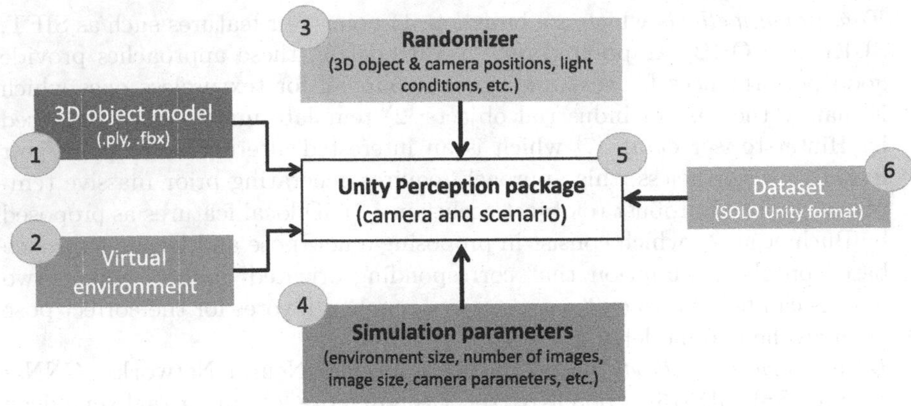

Fig. 1. Framework for creating synthetic labeled datasets

3 Framework for Creating Synthetic Datasets

3.1 Architecture

Figure 1 depicts our framework for creating synthetic labeled datasets. We use the Unity 3D rendering software and its Perception package[1] for creating randomized scenarios made of images and 3D annotated labels. Our framework is made of 6 main stages, describe as follows:

1. *3D object model*: this is the model used in Unity in .ply or .fbx format, with texture.
2. *Virtual environment*: this can be a planar background in front of which the object is placed or this can be a 3D virtual environment in which the object is integrated.
3. *Randomizer*: this is a component in the Unity Perception package which encapsulates specific randomization activities (e.g: positions, orientations, textures, light conditions, etc.) to perform during the execution of a randomized simulation. This randomizer takes the form of a custom script integrating different configurations for each frame of the scenario.
4. *Simulation parameters*: this refers to the number and size of the images, the size of the environment and the camera parameters. Also, it refers to the camera labeler[2].

[1] Unity Perception package is an integrated tool in Unity to realize scenarios with a perception camera that capture frames https://docs.unity3d.com/Packages/com.unity.perception@1.0/manual/index.html.

[2] A labeler is a specific Unity component of the Perception package that marks a GameObject with a specific label. A Labeler is a 2D or 3D bounding box and integrate additional information such as the camera parameters, information about the scene and the object.

5. *Unity Perception*: this block takes in inputs: the 3D model of the object, the virtual environment, the randomizer for generating random scenarios and the simulation parameters.
6. *SOLO Dataset*: this is the output of our framework, a synthetic dataset made of images and 3D object informations, in the Unity SOLO format which consists on: the information of the frame, object position and its orientation regarding the camera, size of the object, the camera parameters required to project the labels on the images.

3.2 Synthetic Labeled Datasets

7 synthetic labeled cardboard box datasets have been created (see Fig. 2). They are organized in 2 categories:

1. Planar background in front of which the object is placed;
2. 3D virtual environment in which the object is inserted with respect to the perspective point of view.

Table 1 depicts the complete list of the datasets and their acronyms.

Table 1. Dataset full descriptive names and associated acronyms.

Planar background datasets	
Fruit	F
Gray	G
Gray with Occlusions	GO
DT Image	DTI
DT Image with Occlusions	DTIO
3D virtual environment datasets	
DT + Object in Pers. Proj.	DTPP
DT + Object in Pers. Proj. and Occlusions	DTPPO

Planar Background Datasets. For creating the planar background datasets, the position of the camera was fixed, a planar background was placed in front of the camera and the positions of the cardboard box has been randomized in between the camera and the planar background positions (see Fig. 3). 5 planar background configurations have been created:

- Gray color with the full object in the camera field of view (Fig. 2a);
- Gray color with occlusions (object partially in the camera field of view) (Fig. 2c);
- Digital Twin image (Fig. 2b);
- Digital Twin image with occlusions (Fig. 2d);
- Fruits randomized (see Fig. 2e).

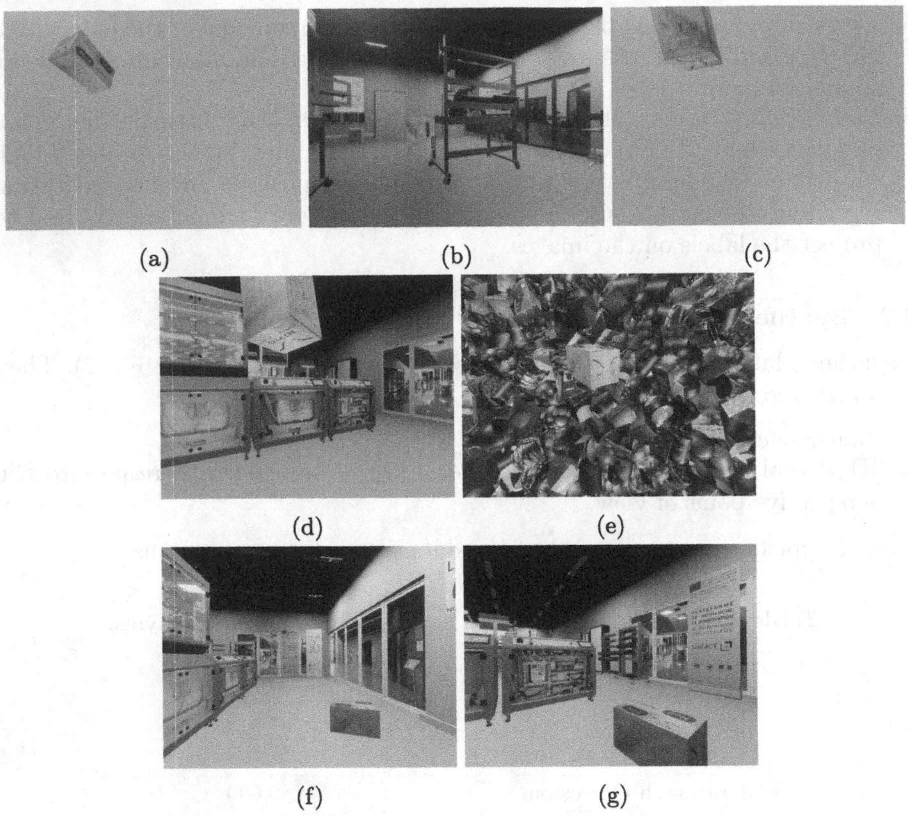

Fig. 2. Dataset image examples. Planar background datasets: (a) - Gray; (b) - DT image; (c) - Gray with occlusions; (d) - DT image with occlusions; (e) - Fruit randomized. 3D virtual environment: (f) - 3D DT; (g) - 3D DT with occlusions.

3D Environment with Object in Perspective Projection Datasets. This category of datasets is built from the DT of the Industry of the Future platform available at CESI campus de Rouen which provides a contextualized 3D background environment in the surrounding of the object. Then, the cardboard box has been randomly integrated in this 3D environment through a perspective projection (cf. Fig. 4). To do so, the camera has been randomly positioned in the DT and a raycasting verifies the presence of an object in its field of view, otherwise the camera is moved to another position. Then, the object was placed in the free space, in the field of view of the camera with a raycasting to check if the cardboard box does not collide with other objects, otherwise the object is moved. In case of occlusions, we manage directly from the images, since controlling the occlusions in 3D is time-consuming and significantly increase the time process for creating the datasets. Finally, 2 datasets have been created:

- 3D DT (see Fig. 2f);
- 3D DT with object occlusions (see Fig. 2g).

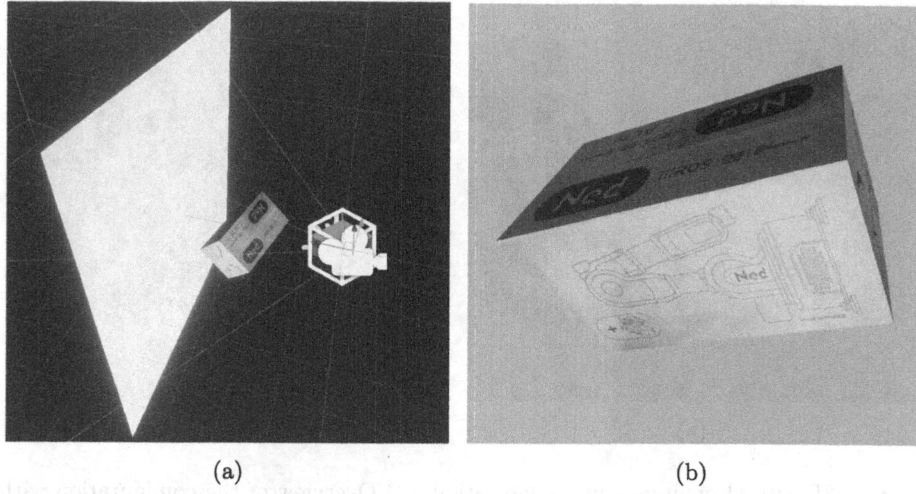

(a) (b)

Fig. 3. Planar background configuration. (a), overview of the Unity scene: the planar background is in white color, the camera is in front of the planar background (represented with the coordinate frame), and in between a cardboard box is inserted. (b), an example of an image grabbed from the camera.

4 Experimental Set-Up

4.1 Implementation Details

To compare the different datasets in the YOLO-6D detection algorithm, the training parameters have been fixes. The size of each dataset was of 5000 images, divided between training and validation as follows: 80% for training (i.e. 4000 images) and 20% validation (i.e. 1000 images). The number of epochs was set to 500. The batch size was fixed to 16. The learning rate was set up to 0.0001. The cardboard dimensions (in m) used for the dataset are: L = 0.56; W = 0.31; H = 0.31.

4.2 Evaluation Metrics

5 metrics have been used for quantifying the performances of each dataset. They are referenced in 2 categories:

3D Metrics

- *Translational Error,* ε_{TE} [9]: ε_{TE} is defined as the L2 norm between the centers, $\overline{\mathbf{P}} = [\overline{\mathbf{R}}, \overline{\mathbf{T}}]$ and $\widehat{\mathbf{P}} = [\widehat{\mathbf{R}}, \widehat{\mathbf{T}}]$, of the 3D bounding boxes, respectively ground-truth and estimated ones (see Eq. 1).

$$\varepsilon_{TE} = ||\overline{\mathbf{T}} - \widehat{\mathbf{T}}||_2 = \sqrt{(\overline{x} - \widehat{x})^2 + (\overline{y} - \widehat{y})^2 + (\overline{z} - \widehat{z})^2} \tag{1}$$

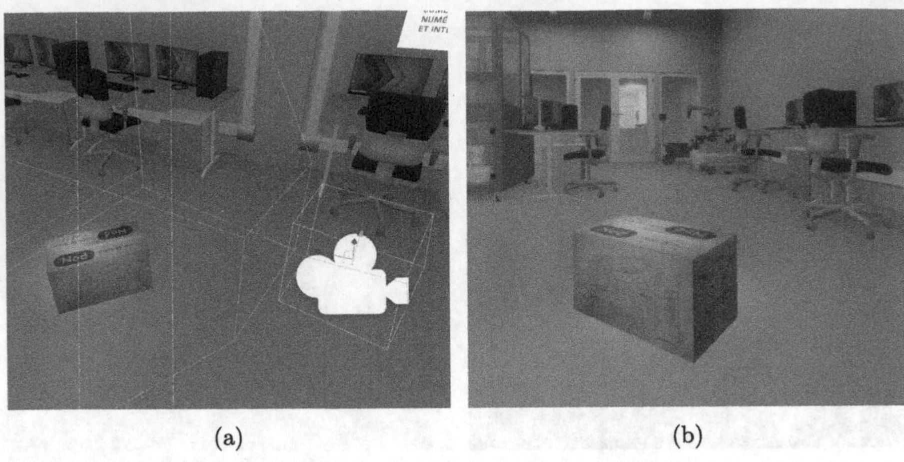

(a) (b)

Fig. 4. 3D virtual environment configuration. (a) Overview of the configuration with a view of the camera and the cardboard box located in the field of view of the camera. (b) an example of an image grabbed from the camera.

- *Rotational Error,* ε_{RE} [9]: ε_{RE} is defined as the angle error between the centers of the 3D ground-truth and predict bounding boxes defined by their rotation matrices 3×3 $\overline{\mathbf{R}}$ and $\widehat{\mathbf{R}}$ (see Eq. (2)).

$$\varepsilon_{RE} = \arccos\left(\frac{\mathrm{Tr}(\widehat{\mathbf{R}}\overline{\mathbf{R}}^{-1}) - 1}{2}\right) \tag{2}$$

- *3D Distance of Center Points,* ε_{DCP}: ε_{DCP} is an extension of the Average 3D distance of model points (ADD) introduced by Kang et al. This is also derived from ε_{TE} since this metric is defined as the distance between the two centers of the bounding boxes: the ground-truth and the estimate. A pose is considered as correct if the distance is less than 10% of the object's diameter.
- *Rotational Accuracy at* 5°, $\varepsilon_{RA_{5^\circ}}$: $\varepsilon_{RA_{5^\circ}}$ is derived from ε_{RE} and defined as the percentage of correctly aligned 3D bounding boxes between ground-truth and the estimate. Here, correctly refers to the rotational error, which must be less or equal to 5° to consider the pose as correct.

2D Metrics

- *Mean Pixel Error,* ϵ_{MPE}: ϵ_{MPE} is defined as the mean distance between the 2D projections of the 3D points of the bounding box using the estimate and the ground-truth pose (Fig. 5).

Fig. 5. Example of results: the 3D bounding box in green refers to the cardboard box ground-truth provided by Unity; the one in red refers to the estimation provided by YOLO-6D. (Color figure online)

5 Results, Analysis and Guidelines

Figures 6 and 7 depict the translational and rotational errors, respectively ε_{TE}, ε_{RE}, through a box plot graph[3]. To achieve these box plot graphs, YOLO-6D has been trained and validated successively on each of the 7 datasets with respect to the following proportions: 80% training and 20% validation. We note that the most convincing results are obtained for the contextualized dataset from the 3D Digital Twin, where the cardboard box has been integrated according to the perspective projection, with occlusions (medians: $\varepsilon_{TE} \approx 0.05\,\text{m}$, $\varepsilon_{RE} \approx 2.5°$) and without occlusions ($\varepsilon_{TE} \approx 0.025\,\text{m}$, $\varepsilon_{RE} \approx 1°$). Planar backgrounds, whether random, uniform color, or an image of the Digital Twin, provide the least relevant

[3] Outliers have been removed to focus the display on the box plot graphs, and thus provide a better visualization of the results.

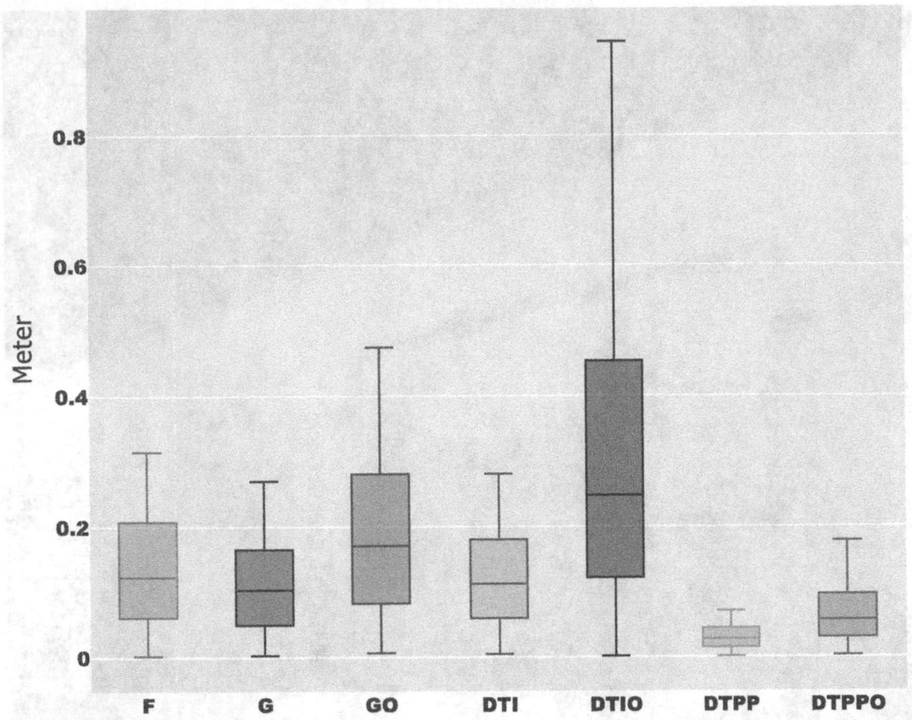

Fig. 6. Translational error (in m) over the datasets between the 3D bounding box centers of the cardboard box: ground-truth provided by Unity and estimation provided by YOLO-6D. See Table 1 for acronym definitions.

results with median translational and rotational errors: $\varepsilon_{TE} \approx [0.1 - 0.25]$ m; $\varepsilon_{RE} \approx [4 - 11]°$.

Tables 2 and 3 depict a list of cross-validation tables carried out over the metrics presented in Sect. 4.2. The proportion for training and validation of YOLO-6D has been unchanged: 80% - training and 20% - validation. The 5 tables must be read as follows: the model has been trained successively on the datasets depicted in line and then, each trained model has been tested on the validation datasets depicted in column. Therefore, the cells cut across the tables, represent the results obtained for a training and a validation stages carried out in the same dataset. Table 2 depicts results for which the most relevant values are the minimum ones, whereas Table 3 depicts results for which the maximum values are the most relevant. Additionally, unlike the box plot graphs which depict medians, numbers in the Tables 2 and 3 refer to average values.

In terms of translational errors (see Table 2a), the most relevant results are obtained for a model trained and validated on the same DTPP dataset ($\varepsilon_{TE} = 0.05$ m). This excellent error value is also achieved for YOLO-6D trained on DTPP dataset and validated on DTPPO dataset. One notice that these results are more competitive than a model trained and validated on DTPPO

Table 2. Cross-validation tables. Minimum values refer to the best results.

(a) Translational Error, ε_{TE}

	F	G	GO	DTI	DTIO	DTPP	DTPPO
F	0.27	3.65	3.95	1.7	3.75	3.87	3.37
G	1.52	0.21	0.24	0.23	1.36	1.51	1.44
GO	1.79	0.52	0.38	0.51	9.94	0.51	9.94
DTI	0.88	6.25	4.47	0.28	1.34	4.65	4.00
DTIO	0.97	5.56	4.55	0.84	1.46	5.26	4.33
DTPP	2.17	2.70	3.36	3.30	9.45	0.05	0.05
DTPPO	8.45	5.20	5.53	5.83	11.97	0.38	0.21

(b) Rotational Error, ε_{RE}

	F	G	GO	DTI	DTIO	DTPP	DTPPO
F	5.25	120	115	37	80	125	125
G	36.4	8.18	8.91	9.36	38	120	122
GO	45	16	13.7	18	42	18.6	42
DTI	19	72	63	9.12	30	120	122
DTIO	26.8	92	84	20	33.1	122	123
DTPP	88	49	52	42	93	2.51	3.36
DTPPO	107	93.6	88	61	110	7.22	5.36

(c) Mean Pixel Error, ϵ_{MPE}

	F	G	GO	DTI	DTIO	DTPP	DTPPO
F	12	502	481	108	264	477	506
G	145	8	11	8	98	124	141
GO	215	30	30	38	149	38	149
DTI	40	275	238	13	72	314	408
DTIO	80	436	311	51	78	335	460
DTPP	305	169	192	150	452	3	4
DTPPO	385	342	353	252	591	15	5

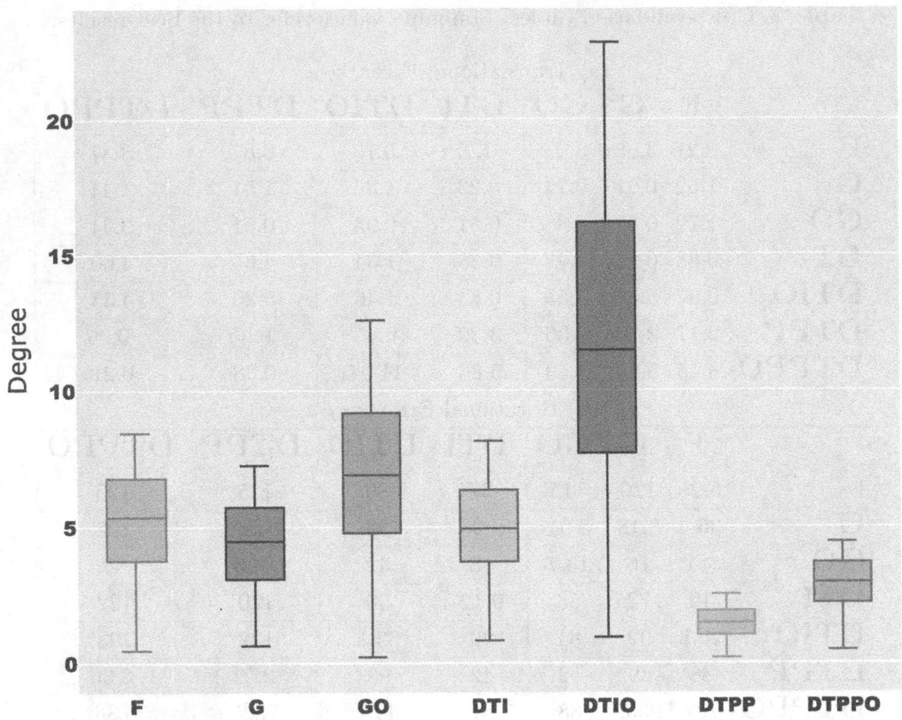

Fig. 7. Rotational error (in °) over the datasets between the 3D bounding box centers of the cardboard box: ground-truth provided by Unity and estimation provided by YOLO-6D.

dataset ($\varepsilon_{TE} = 0.21$ m). This highlights that the YOLO-6D model trained with DTPP dataset is robust in the presence of data close to real-life situations (i.e. with occlusions).

In terms of rotational errors, Table 3b highlights different tendencies since planar backgrounds get the most relevant results, specifically when DTI dataset is used for training the model and DTIO is used as the validation dataset ($\varepsilon_{RE} = 0.08°$). Results for DTPP and DTPPO datasets are the least convincing, since ε_{RE} are x10 higher: ε_{RE}(DTPP, DTPPO) $= [0.80, 0.70, 0.81, 0.78]°$.

Tables 2b, 2c, 3 depict the sames tendencies as Table 2a since the most convincing results are obtained for YOLO-6D trained and validated on DTPP dataset: $\varepsilon_{DCP} = 69\%$; $\varepsilon_{RA_{5°}} = 2.51$; $\epsilon_{MPE} = 3$. Also, we note the same tendencies as in Table 2a since competitive results are obtained for YOLO-6D trained on DTPP and validated on DTPPO datasets: ($\varepsilon_{DCP} = 59\%$; $\varepsilon_{RA_{5°}} = 3.36$; $\epsilon_{MPE} = 4$). This confirms the robustness of the model trained with the DTPP dataset, providing very encouraging results for real-life situations where the cardboard box can be partially occluded.

Table 3. Cross-validation table. Values are in %. Maximum values refer to the best results.

(a) 3D Distance of Center Points, ε_{DCP}

	F	G	GO	DTI	DTIO	DTPP	DTPPO
F	14	0	0	60	1	0	0
G	1	20	13	17	1	0	0
GO	1	13	10	12	1	12	1
DTI	30	1	1	16	20	0	0
DTIO	4	1	1	17	40	0	0
DTPP	0	3	5	2	0	69	59
DTPPO	0	0	10	1	0	41	40

(b) Accuracy at 5°, $\varepsilon_{RA_{5^\circ}}$

	F	G	GO	DTI	DTIO	DTPP	DTPPO
F	54	0	0	20	1	0	0
G	1	69	50	60	4	1	0
GO	1	48	38	40	2	40	2
DTI	13	6	3	64	8	0	0
DTIO	19	3	1	54	12	0	0
DTPP	0	1	1	1	0	80	70
DTPPO	0	0	0	3	0	81	78

This also suggests that it is not necessary to include every possible situation by adding images of partially masked boxes to the training dataset: the results tend to depict that this is counterproductive, with degraded performances.

6 Conclusion and Perspectives

In this paper, we have introduced a framework for creating synthetic datasets for industrial object detection and 6D pose estimation based on Unity 3D rendering software. The key-element relies on the Unity Perception package, which allows the automatic data labeling for producing instantaneously massive annotated datasets. In addition, we have benchmarked several synthetic datasets for industrial cardboard box detection and pose estimation based on YOLO-6D. Results highlight some guidelines for creating industrial object datasets: 1) the relevance of training the DL model with 3D contextualized synthetic environments in which objects are integrated with respect to the perspective projection; 2) including images with occlusion in the training dataset, to make it as representative as possible of real situations, degrades detection and pose estimation performances.

Our future works will aim to increase the datasets by adding new industrial objects. Next, we will evaluate the deep learning model trained with synthetic labeled images, in real situations. This set-up will be carried out in the Industry of the Future platform of CESI laboratory (https://lineact.cesi.fr/en/factory-of-the-future/) with the use of lab capabilities: MIR robot to grab the images; motion capture system to get the ground-truth pose of the cardboard box. Finally, we will look at the influence of the number of real vs. synthetic images on detection and 6D pose estimation performances of industrial objects.

Acknowledgements. This work has been carried out within the framework of the COLIBRY project "COLlaborative semantIc roBotics for industRY 5.0" (https://lineact.cesi.fr/projet-colibry/). COLIBRY is a prizewinner of the Metropole Rouen Normandie's 2021 "Research" call for projects, accredited by the Cosmetic Valley competitiveness cluster and supported by the Normandie AeroEspace sector.

References

1. Brachmann, E., Krull, A., Michel, F., Gumhold, S., Shotton, J., Rother, C.: Learning 6D object pose estimation using 3D object coordinates. In: Fleet, D., Pajdla, T., Schiele, B., Tuytelaars, T. (eds.) ECCV 2014. LNCS, vol. 8690, pp. 536–551. Springer, Cham (2014). https://doi.org/10.1007/978-3-319-10605-2_35
2. Buch, A.G., Kiforenko, L., Kraft, D.: Rotational subgroup voting and pose clustering for robust 3D object recognition, pp. 4137–4145. IEEE Computer Society (2017). ISSN 2380-7504
3. Calli, B., et al.: Yale-CMU-Berkeley dataset for robotic manipulation research. Int. J. Rob. Res. **36**(3), 261–268 (2017)
4. Chen, J., Zhang, L., Liu, Y., Xu, C.: Survey on 6D pose estimation of rigid object. In: 2020 39th Chinese Control Conference (CCC), pp. 7440–7445 (2020). ISSN 1934-1768
5. De Roovere, P., Moonen, S., Michiels, N., Wyffels, F.: Dataset of industrial metal objects (2022). arXiv:2208.04052 [cs]
6. Do, T.-T., Cai, M., Pham, T., Reid, I.: Deep-6DPose: recovering 6D object pose from a single RGB image (2018). _eprint: 1802.10367
7. Hinterstoisser, S., et al.: Multimodal templates for real-time detection of texture-less objects in heavily cluttered scenes. In: 2011 International Conference on Computer Vision, pp. 858–865 (2011). ISSN 2380-7504
8. Hinterstoisser, S., et al.: Model based training, detection and pose estimation of texture-less 3D objects in heavily cluttered scenes. In: Lee, K.M., Matsushita, Y., Rehg, J.M., Hu, Z. (eds.) ACCV 2012. LNCS, vol. 7724, pp. 548–562. Springer, Heidelberg (2013). https://doi.org/10.1007/978-3-642-37331-2_42
9. Hodaň, T., Matas, J., Obdržálek, Š: On evaluation of 6D object pose estimation. In: Hua, G., Jégou, H. (eds.) ECCV 2016. LNCS, vol. 9915, pp. 606–619. Springer, Cham (2016). https://doi.org/10.1007/978-3-319-49409-8_52
10. Hodaň, T., Haluza, P., Obdržálek, Š., Matas, J., Lourakis, M., Zabulis, X.: T-LESS: an RGB-D dataset for 6D pose estimation of texture-less objects. In: IEEE Winter Conference on Applications of Computer Vision (WACV) (2017)
11. Huang, Z., Shen, Y., Li, J., Fey, M., Brecher, C.: A survey on AI-driven digital twins in industry 4.0: smart manufacturing and advanced robotics. Sensors **21**(19), 6340 (2021)

12. Kang, J., Liu, W., Tu, W., Yang, L.: YOLO-6D+: single shot 6D pose estimation using privileged silhouette information. In: 2020 International Conference on Image Processing and Robotics (ICIP), pp. 1–6 (2020)
13. Kehl, W., Manhardt, F., Tombari, F., Ilic, S., Navab, N.: SSD-6D: making RGB-based 3D detection and 6D pose estimation great again, pp. 1530–1538. IEEE Computer Society (2017). ISSN 2380-7504
14. Ljungqvist, M.G., Nordander, O., Skans, M., Mildner, A., Liu, T., Nugues, P.: Object detector differences when using synthetic and real training data. SN Comput. Sci. **4**(3), 302 (2023)
15. Marchand, E., Uchiyama, H., Spindler, F.: Pose estimation for augmented reality: a hands-on survey. IEEE Trans. Vis. Comput. Graph. **22**(12), 2633–2651 (2016)
16. Pitteri, G.: Estimation de la pose 3D d'objets dans un environment industriel. Theses, Université de Bordeaux (2020). Issue: 2020BORD0202
17. Rad, M., Lepetit, V.: BB8: a scalable, accurate, robust to partial occlusion method for predicting the 3D poses of challenging objects without using depth, pp. 3848–3856. IEEE Computer Society (2017). ISSN 2380-7504
18. Tekin, B., Sinha, S.N., Fua, P.: Real-time seamless single shot 6D object pose prediction, pp. 292–301. IEEE Computer Society (2018)
19. Vanherle, B., Moonen, S., Van Reeth, F., Michiels, N.: Analysis of training object detection models with synthetic data (2022). _eprint: 2211.16066
20. Xiang, Y., Schmidt, T., Narayanan, V., Fox, D.: PoseCNN: a convolutional neural network for 6d object pose estimation in cluttered scenes. In: Robotics: Science and Systems (RSS) (2018)
21. Zhang, X., Jiang, Z., Zhang, H.: Real-time 6D pose estimation from a single RGB image. Image Vis. Comput. **89**, 1–11 (2019)

Author Index

S.-H. Sheu (Ed.): ICIEA-EU 2024, LNBIP 507, pp. 243–244, 2024.
https://doi.org/10.1007/978-3-031-58113-7

Printed in the United States
by Baker & Taylor Publisher Services

Printed in the United States
by Baker & Taylor Publisher Services